Managerial Approaches Toward Queuing Systems and Simulations

Salvador Hernandez-Gonzalez
Technological Institute of Celaya, Mexico

Manuel Dario Hernandez Ripalda
Technological Institute of Celaya, Mexico

A volume in the Advances in Mechatronics and Mechanical Engineering (AMME) Book Series

IGI Global
DISSEMINATOR OF KNOWLEDGE

Published in the United States of America by
IGI Global
Engineering Science Reference (an imprint of IGI Global)
701 E. Chocolate Avenue
Hershey PA, USA 17033
Tel: 717-533-8845
Fax: 717-533-8661
E-mail: cust@igi-global.com
Web site: http://www.igi-global.com

Library of Congress Cataloging-in-Publication Data

Names: Hernandez-Gonzalez, Salvador, 1973- author. | Hernandez Ripalda, Manuel Dario, 1954- author.
Title: Managerial approaches toward queuing systems and simulations / by Salvador Hernandez-Gonzalez and Manuel Dario Hernandez Ripalda.
Description: Hershey, PA : Engineering Science Reference, [2018] | Includes bibliographical references.
Identifiers: LCCN 2017038308| ISBN 9781522552642 (hardcover) | ISBN 9781522552659 (ebook)
Subjects: LCSH: Queuing theory--Mathematics. | Stochastic systems--Mathematical models.
Classification: LCC T57.9 .H454 2018 | DDC 519.8/5--dc23 LC record available at https://lccn.loc.gov/2017038308

This book is published in the IGI Global book series Advances in Mechatronics and Mechanical Engineering (AMME) (ISSN: 2328-8205; eISSN: 2328-823X)

British Cataloguing in Publication Data
A Cataloguing in Publication record for this book is available from the British Library.

Advances in Mechatronics and Mechanical Engineering (AMME) Book Series

ISSN:2328-8205
EISSN:2328-823X

Editor-in-Chief: J. Paulo Davim, University of Aveiro, Portugal

MISSION

With its aid in the creation of smartphones, cars, medical imaging devices, and manufacturing tools, the mechatronics engineering field is in high demand. Mechatronics aims to combine the principles of mechanical, computer, and electrical engineering together to bridge the gap of communication between the different disciplines.

The **Advances in Mechatronics and Mechanical Engineering (AMME) Book Series** provides innovative research and practical developments in the field of mechatronics and mechanical engineering. This series covers a wide variety of application areas in electrical engineering, mechanical engineering, computer and software engineering; essential for academics, practitioners, researchers, and industry leaders.

COVERAGE

- Computer-Based Manufacturing
- Design and Manufacture
- Biologically Inspired Robotics
- Control Systems Modelling and Analysis
- Computational Mechanics
- Tribology and surface engineering
- Nanomaterials and nanomanufacturing
- Mechanisms and machines
- Sustainable and green manufacturing
- Autonomous systems

IGI Global is currently accepting manuscripts for publication within this series. To submit a proposal for a volume in this series, please contact our Acquisition Editors at Acquisitions@igi-global.com or visit: http://www.igi-global.com/publish/.

The Advances in Mechatronics and Mechanical Engineering (AMME) Book Series (ISSN 2328-8205) is published by IGI Global, 701 E. Chocolate Avenue, Hershey, PA 17033-1240, USA, www.igi-global.com. This series is composed of titles available for purchase individually; each title is edited to be contextually exclusive from any other title within the series. For pricing and ordering information please visit http://www.igi-global.com/book-series/advances-mechatronics-mechanical-engineering/73808. Postmaster: Send all address changes to above address.

Titles in this Series

For a list of additional titles in this series, please visit:
https://www.igi-global.com/book-series/advances-mechatronics-mechanical-engineering/73808

Advanced Numerical Simulations in Mechanical Engineering
Ashwani Kumar (Government of Uttar Pradesh, India) Pravin P. Patil (Graphic Era University, India) and Yogesh Kr. Prajapati (National Institute of Technology Uttarakhand,India)
Engineering Science Reference • ©2018 • 242pp • H/C (ISBN: 9781522537229) • US $165.00

Numerical and Analytical Solutions for Solving Nonlinear Equations in Heat Transfer
Davood Domiri Ganji (Babol Noshirvani University of Technology, Iran) and Roghayeh Abbasi Talarposhti (University of Mazandaran Iran)
Engineering Science Reference • ©2018 • 275pp • H/C (ISBN: 9781522527138) • US $195.00

Nanofluid Technologies and Thermal Convection Techniques
Ramesh Chand (Government Arya Degree College Nurpur (HP), India)
Engineering Science Reference • ©2017 • 229pp • H/C (ISBN: 9781683180067) • US $185.00

Mathematical Concepts and Applications in Mechanical Engineering and Mechatronics
Mangey Ram (Graphic Era University, India) and J. Paulo Davim (University of Aveiro, Portugal)
Engineering Science Reference • ©2017 • 495pp • H/C (ISBN: 9781522516392) • US $215.00

Handbook of Research on Advances and Applications in Refrigeration Systems and...
Pedro Dinis Gaspar (University of Beira Interior, Portugal) and Pedro Dinho da Silva (University of Beira Interior, Portugal)
Engineering Science Reference • ©2015 • 924pp • H/C (ISBN: 9781466683983) • US $545.00

Engineering Creative Design in Robotics and Mechatronics
Maki K. Habib (The American University in Cairo, Egypt) and J. Paulo Davim (University of Aveiro, Portugal)
Engineering Science Reference • ©2013 • 366pp • H/C (ISBN: 9781466642256) • US $195.00

For an entire list of titles in this series, please visit:
https://www.igi-global.com/book-series/advances-mechatronics-mechanical-engineering/73808

IGI Global
DISSEMINATOR OF KNOWLEDGE

701 East Chocolate Avenue, Hershey, PA 17033, USA
Tel: 717-533-8845 x100 • Fax: 717-533-8661
E-Mail: cust@igi-global.com • www.igi-global.com

Table of Contents

Preface

The person in charge of a process (physician, engineers, managers) needs a fundamental knowledge base; formed by concepts and basic principles, that provide quantitative laws that allow her/him to know how to approach the operational problems of the systems under her/his responsibility; from a rational basis of understanding and search; contributing the ability to evaluate and improve systems and processes.

This book is a support in the formation of the ability to take decisions through the analysis of issues and difficulties that arise in productive or service systems; subject to the effects of randomness (e.g. on demand or service time); in which it is necessary to estimate the time of permanence in the system; the amount of work in progress; calculate flows, costs and routes; considering quantitative and economic criteria as measures of performance.

It should be noted that the analysis of systems subjected to the effects of randomness has been an area covered so far by the simulation texts; the main reason has been the lack of analytical models of the systems or due to the computational effort required to obtain a solution. Thanks to the advances of the last 20 years in the development of analytical models this has been reversed and in our book the same problems are now analyzed with an approach where the use of analytical models predominates; these models are taken from queueing theory.

One good thing about analytical models is that a spreadsheet is enough to apply them; if the model is more complex then one can resort to a programming language (Many engineers have fun writing programs); there are also special purpose programs for the management of resources that carry out transactions. Combining the use of analytical models with a simulation model greatly reinforces the understanding of a system.

There is abundant literature on queueing theory and its application for the analysis of systems subject to randomness: Kleinrock (1975), Cooper (1981), Allen (1990), Buzacott and Shanthikumar (1993), and Altiok (1997) are just some fundamental references on the subject. Other more recent books are those of Balsamo, Nitto-Persone, and Onvural (2001), Bolch, Greiner, de Meer, Trivedi (2006), and Gross, Shortle, and Thompson (2008). In all of them the topic of queueing theory is treated with an emphasis on the aspect of reasoning and demonstration of each one of the models. But decision-making is also necessary.

The books of Operations Research and Operations Management are those that have covered the application to the decision making in the systems; however, its scope tends to be general: Render and Stair (1997), Taha (2011), Hillier and Lieberman (2005) are some basic references.

The texts of Hall (1991), Hopp and Spearman (2008), and Curry and Feldman (2011) are focused on analyzing the problems of systems using analytical models provided by queueing theory. Our text follows this line of work.

The book is for students of management and industrial engineering as well as practitioners in general; which have under their responsibility the administration of systems formed by several stages; who need a book that addresses the problems of modeling systems and manufacturing services subject to the effects of randomness and also provides them with a theoretical basis that allows them to understand how the elements that are part of a system are related. We use spreadsheets and R language packages to perform several of the calculations shown in the examples; given that we are aware that many professionals have programming skills; we also present some codes in Scilab.

The present book was developed while both authors were working in Tecnológico Nacional de México en Celaya.

REFERENCES

Allen, A. O. (1990). *Probability, statistics, and queueing theory: With computer science applications* (2nd ed.). New York: Academic Press.

Altiok, T. (1997). *Performance analysis of manufacturing systems*. New York: Springer. doi:10.1007/978-1-4612-1924-8

Balsamo, S., Nitto-Persone, V., & Onvural, R. (2001). *Analysis of Queueing Networks with Blocking. New Yor*. Springer. doi:10.1007/978-1-4757-3345-7

Bolch, G., Greiner, S., de Meer, H., & Trivedi, K. S. (2006). *Queueing networks and Markov chains*. Wiley. doi:10.1002/0471791571

Buzacott, J. A., & Shanthikumar, J. G. (1993). *Stochastic models of manufacturing systems*. Upper Saddle River, NJ: Prentice Hall.

Cooper, R. B. (1981). *Introduction to Queueing Theory* (2nd ed.). New York: Elsevier North Holland.

Curry, G. L., & Feldman, R. M. (2011). *Manufacturing systems modeling and analysis* (2nd ed.). Berlin: Springer. doi:10.1007/978-3-642-16618-1

Gross, D., Shortle, J. F., & Thompson, J. M. (2008). *Fundamentals of queueing theory*. Hoboken, NJ: J. Wiley & Sons. doi:10.1002/9781118625651

Hall, R. W. (1991). *Queueing methods: For services and manufacturing*. Englewood Cliffs, NJ: Prentice Hall.

Hillier, F. S., & Lieberman, G. J. (2005). *Introduction to operations research* (8th ed.). New York, NY: McGraw-Hill.

Hopp, W. J., & Spearman, M. L. (2008). *Factory physics*. Long Grove, IL: Waveland Press.

Kleinrock, L. (1975). *Queueing systems*. New York, NY: Wiley.

R Development Core Team. (2008). *R: A language and environment for statistical computing*. R Foundation for Statistical Computing. Retrieved from http://www.R-project.org

Render, B., & Stair, R. M. (1997). *Quantitative analysis for management*. Upper Saddle River, NJ: Prentice Hall.

Scilab Enterprises. (2012). Scilab: Free and Open Source software for numerical computation (OS, Version 5.3.3) [Software]. Available from: http://www.scilab.org

Taha, H. A. (2011). *Operations research: An introduction* (9th ed.). Upper Saddle River, NJ: Prentice Hall.

Chapter 1
Quantitative Analysis

ABSTRACT

The first chapter presents concepts about quantitative models for decision making; the process that follows is used when constructing a model and definitions of queueing theory: work in process, cycle time, and congestion. Also presented and explained is the Kendall notation that will be used throughout the book. The chapter shows some examples of waiting line systems at the end.

QUANTITATIVE ANALYSIS

Quantitative analysis is defined as the scientific approach to problems or conflictive situations in a system, where the administrator has to make a decision.

Systems consist of a variety of components or elements. Each element is connected to the others and together they form an entity that produces an answer. Systems, in decision-making, are those entities that are under the responsibility of an administrator who has to make a series of decisions that should generate the answer needed to achieve an objective.

The systems of interest are organizations with production and distribution targets. Some examples of systems that have only a few components are grocery stores, cafeterias, stationery stores, small supermarkets, small restaurants, bakeries. In these cases the managers are often the owners themselves and are in charge of all the business' planning, buying, contracting, selling, accounting and dispatching jobs. Given that these systems do not have a lot of personnel, the quantity of consumables is small and the variety of products on offer is low, they are usually relatively simple to manage.

DOI: 10.4018/978-1-5225-5264-2.ch001

Some examples of systems with a lot of components are car manufacturers, airports, supermarkets, universities, airlines, online sales companies. In these cases, the systems have a huge variety of components, each of which involves several people. There will often be someone in charge of each area, which shall, in turn, be coordinated by an entity consisting of a single person or by a board composed of several officers with specialist duties. In these systems, decisions may be complex.

In quantitative analysis there are a set of tools that reduce the uncertainty associated with decision-making. These tools are often mathematical models that enable us visualize how each of the components of the system (variables) is connected to the others.

Models

Models are representations of systems and describe the characteristics that are of interest for the decisions that the administrator has to make. The process of creating a model is known as *modeling*. Models can be any of three types: physical, mental or symbolic.

- **Physical Models:** Physical models are scale models of airplanes, houses or cities, for example. This way of representing systems was one of the first to be applied for system analysis. During the two World Wars, people used scale representations of the battlefields. These models were built with the detailed location of the trenches, gunner positions, roads, cities, together with the location of each of the divisions that made up the armies or of the squadrons of enemy fighter planes and bombers.
- **Mental:** This is one of the first representations of systems ever used and is called the imagination. As children we have all used this representation intensively.
- **Symbolic:** A symbolic model shows the relationships between each component of a system through the use of symbols or operators. The best known ones are mathematical models. A supply equation is a symbolic representation of the behavior of demand in respect of the price of an item or product. A queueing formula is a representation of the behavior of the cycle time in respect of the demand for a service (for example, in a cafeteria) and the service capacity of the server (the cashier who takes the order and charges).

Modeling Stages

A mathematical model is not built by chance. Mathematical models are built out of the need to represent a system; in the field of decision-making, mathematical models are used to speed up decision-making and reduce the uncertainty associated with the effect or result of taking an action.

A series of stages or phases are named and described below (Figure 1).

Definition of the Problem

As has already been mentioned, the idea of developing a model arises from the need to solve a problem or conflict (as posed here) relating to the administration of a system. Defining a problem is (often) the most complicated part. The better the lack or need is defined, the better defined the resulting model will be.

Construction of the Model

Once the problem or situation to be modeled has been identified, the person in charge should commence with a process of analysis that will lead them to a representation of the reality of the situation. The process may, for example, be carried out using logical mathematical reasoning to get an equation.

It is important to point out that it is the convention in model building to start out with simplified systems with not much complexity or with small systems, only taking a few variables into account, and then, as soon as a satisfactory result has been obtained, new elements can be incorporated to enlarge the model.

Figure 1. Process modeling

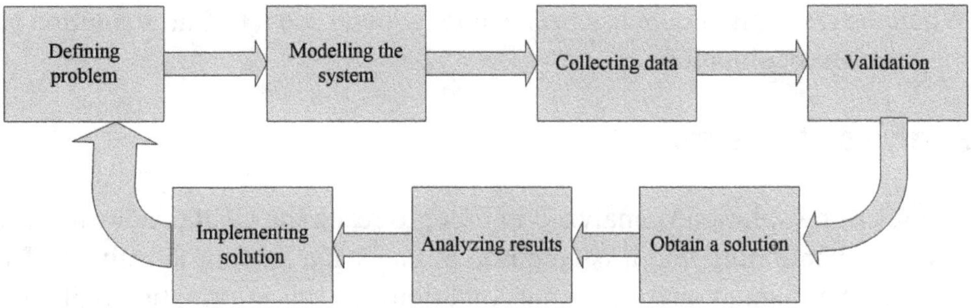

Developing a model of very complex systems or taking all the variables into account is usually extremely complicated, particularly if the necessary experience with the system is lacking. This process can easily lead to confusion.

Data Collection

This stage consists the administrator collecting the available information either from past records or a database. It is worth mentioning that it is normal to take a specific data sample to corroborate the information available in the records. The data are processed, which, for example, implies obtaining the data relating to descriptive statistics, such as the mean, standard deviation, histograms and sometimes carrying out goodness-of-fit tests.

To collect the data, the conditions must be recorded under which the observation was made, such as the place, time of day, identifying components, identifying the object to be sampled. This is necessary because, should you wish to carry out a review, having your records properly identified and in order means you will be able to quickly identify any problem or mistake.

Model Validation

As the name of this stage implies, once you have the model, you need to carry out tests by making controlled predictions under known conditions and comparing them with the real-life situation. If the model does not exist or it is hard to perform a test with the real system, you may opt for a simulation, experts' opinions or testing the model under extreme situations.

If there are doubts about the results, then the model should be reviewed in order to debug it and make the pertinent changes.

Solving the Model

When the resulting model has been validated, you use it to find a solution to the problem or situation.

Analysis of Results

The resulting solution is analyzed in order to reach the right conclusions. It is worth mentioning that it is common in this stage to want to estimate the effect of the changes on the original conditions. So this information is altered in the model for this purpose and the proper prediction obtained. From the analysis we can estimate which variable has a greater effect on the system or identify scenarios where changes in the conditions lead to totally different results from the ones that intuition and experience would have led us to expect.

Implementation

Once a solution has been found for the problem, an answer obtained using the model and the results analyzed, it is time to make the decision to implement the answer that most suits the administrator's interests or objectives. It is worth mentioning that one person is not usually solely responsible for the implementation but rather a group or team are involved, so the implementation has to be well thought through in order to make sure that it is what was really wanted.

The process described above is neither immediate nor linear, but depends, above all, on the complexity of the problem, the availability of information or on the existence of background information. Usually, the person or people in charge of the modeling do not start from nothing but rather consult a variety of sources of information to identify other models that have been developed and applied to similar situations. In this case, the modeling process focuses on adapting an existing model to the present conditions.

Although this work is time consuming, it is generally found to be worth it.

The Phenomenon of Waiting

As we go about our daily lives, we often have to go somewhere to ask for a service: a cafeteria, a restaurant, a movie theater, a doctor's office or to get official documents at a government office. When we arrive we usually find that the person providing the service is busy looking after someone else and we have to wait our turn.

We have a similar situation in industry: pieces arrive at the machines where they are to undergo a manufacturing process, which does not happen in an instant, they take time. If a new piece arrives at the machine while the operation is being performed, then it must wait its turn.

Waiting is a phenomenon that is inherent to processes and basically its perception is based on time. In today's globalized world where companies compete to provide the best service to their customers, the time spent waiting is one of several measurements used to judge the quality of a service.

How many times do we have to wait for 1 or 2 hours in a government office or at the bank to do 15 minutes' worth of paperwork?

Or when it is finally our turn to see the doctor, the doctor spends 30 minutes taking our vital signs, asking questions, making a diagnosis and finally writing out and giving us the respective prescription?

How many times have we decided to leave a line when we see that the cost-benefit of waiting is not in our favor?

When we have to cross a big city at rush hour, we find the roads are saturated and the traffic is slow, so it takes us longer to reach our destination. If, on top of that, the traffic lights at the street crossings are badly synchronized or there is an accident, the situation is even worse.

The managers of some supermarkets take the phenomenon of waiting very seriously and nowadays we see more and more posters with this type:

If you are customer number 5 in the line, please tell our supervisor and we will open up a new till immediately.

Some cities also make some decisions aimed at speeding up the flow of traffic, such as building bridges or changing the direction of some roads. This should mean that people spend less time in their cars.

The waiting phenomenon is studied by a specialist field called Queueing Theory. Over the years, this field has developed a set of models that enable us to understand the causes of waiting. Queueing Theory measures its impact in quantitative terms and analyzes different alternatives or options speed up the flow of customers through a system.

Elements of a Queueing System

The object of study is known as a Queueing system, which is made up by a set of elements (Figure 2).

- **Customer:** A patient or a computer user in an internet café. In general, this is the entity that asks for a service and should be served. The customers are sometimes not obvious or evident at first sight, for example, a passenger waiting at a bus stop or terminal.

The customers may arrive to request a service either randomly or at well defined intervals of time (typical of automated processes). The entry of a customer is called an arrival.

Figure 2. A queueing system: clients, server, queue, and demand

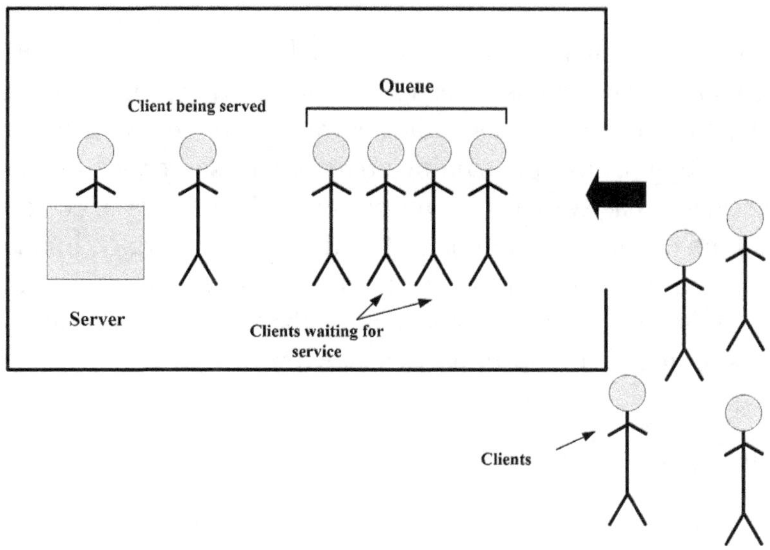

- **Server:** This is the person who serves the customer a coffee and charges for it; the machine that processes the piece; or the machine that collects parking fees. Queueing systems may have a single server or several parallel servers. The server requires a certain time to execute a task (the service or manufacturing process) and this usually varies a little from customer and customer. The main differences in time should be:
  - The quantity of goods the customer buys in a supermarket
  - The level of emergency of a patient in a hospital
  - The number of signatures a document needs.

As the server does not have any control over the characteristics of the customers that arrive, the service time too is usually random. Likewise, the discipline or policy for attending to users depends on the type of system.

- **Queue:** This is made up by to be customers waiting to be served. The length of a queue makes a strong visual impact and people make a direct connection between its size and the quality of service, seeing it as the time it takes to be served. If the queue is "very long" the customer's perception is one of bad and chaotic service, so a user could be easily discouraged and may balk at the idea of joining the queue.

Service Capacity

As we have already mentioned, queueing systems are inherent to processing systems, neither serving a customer nor a manufacturing process are something that can be done in the instant. On the contrary, it all requires a certain length of time, resulting in customers piling up before the server, waiting their turn.

The service capacity of a queueing system is the number of customers that can be served during a certain period of time (t_s). Generally, in queues, the average number of customers served per unit of time is used as a measure of the capacity of a server and represented by the Greek letter μ (mu). Service capacity is in inverse proportion to service time:

$$\text{Attention Capacity} = \mu = \frac{1}{t_S} \tag{1}$$

In other words, if each piece requires about 5 minutes per piece, then applying (1), the service capacity is 1 customer every 5 minutes or, converting it into units of hours:

$$\text{Attention Capacity} = \mu = \frac{1}{5}\frac{\text{customers}}{\text{minute}} \times \left(\frac{60}{1}\frac{\text{minute}}{\text{hour}}\right) = 12\frac{\text{customers}}{\text{hour}}$$

Arrivals Process

The businesses in the service area receive a certain number of customers demanding service every day. Customers or pieces may arrive one at a time or in groups (called batches) and a batch be made up by a set or a random number of customers. The pieces are sent to the equipment to be processed at a certain rhythm and, in general, customers arrive at the system with a certain periodicity (t_a).

In queues, the number of customers that arrive per time period is represented by the Greek letter λ (lambda) and is expressed as:

$$\text{Demand} = \lambda = \frac{1}{t_a} \tag{2}$$

For example, if 1 customer arrives at a system every 10 minutes, then by applying (2):

$$\text{Demand} = \lambda = \frac{1}{10}\frac{\text{cutomers}}{\text{minute}} \times \left(\frac{60}{1}\frac{\text{minute}}{\text{hour}}\right) = 6\frac{\text{customers}}{\text{hour}}$$

Service Discipline

This refers to the order or priority in which customers or users are served. The most common one is First In, First Out. This discipline is the one used by most systems and implies that there is no rule or principle that indicates that any customer should have preference over another. But we must consider that there are other customer service policies too (Table 1).

In hospitals, for example, it is common practice to assign priorities for attention to the patients based on the risk to the patient's life. There are patients that arrive with an injury are illness that puts their life at risk and cannot be made to wait to be seen, whereas, on the other hand, there are patients with symptoms that indicate that their life is not at risk and can wait to be assigned a doctor for their care (Figure 3).

Classification of Queues

A complete service often implies going through several stages or phases, so it is very unusual for the service to be completed in a single stage. Queueing systems can also be classified according to the stages that they should go through during the transformation process.

Table 1. Attention policies

Attention Discipline	Symbol	Examples
First In First Out	FIFO	Most of systems like Banks, stores, restaurants, parking lots
Last-In-First-Out	LIFO	Elevator
Service In – Random Order	SIRO	Pieces in boxes or containers in manufacturing
Priority	P	Hospitals

Figure 3. A queue system with priorities. The patient with the highest level of priority will enter to be served by the doctor, no matter the amount of time elapsed in the queue.

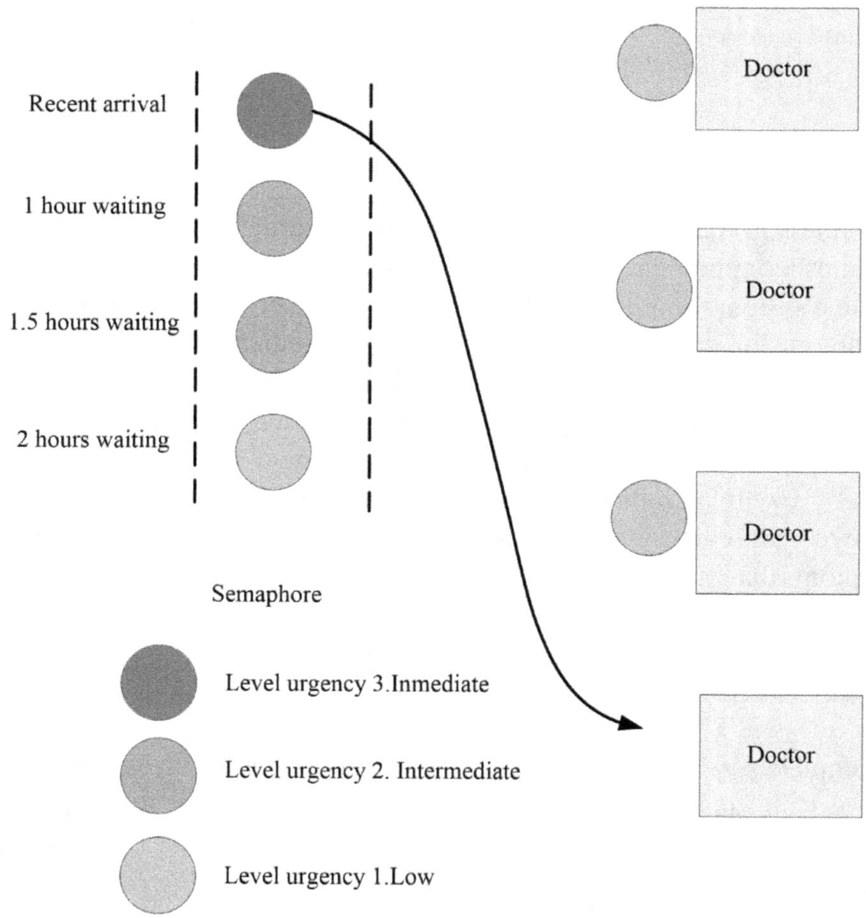

Single-Stage, Single-Server System

This is the simplest kind. The customer enters the system where they wait to be served, then as soon as a server is free, the customer goes in be served. When the service finishes, the customer leaves the system and the next customer goes in (Figure 4).

These systems are typical of small businesses such as ice-cream parlors, coffee shops and small grocery stores.

Figure 4. Single stage queue

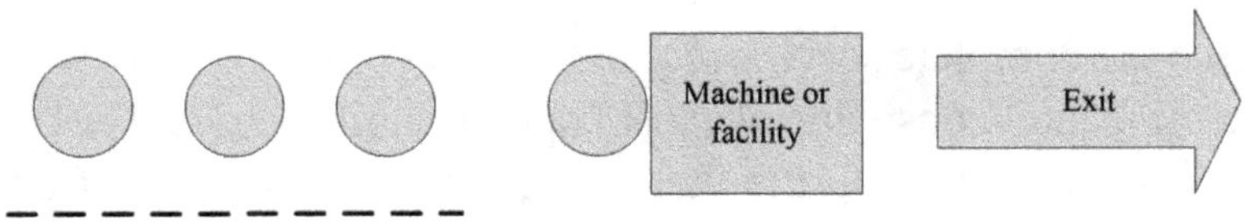

Single-Stage, Parallel Servers Systems

As in the above example, the service is completed in just one stage, however in this case there are a variety of servers serving several customers at the same time. Once a server finalizes the attention of his customer, it is free to receive the next waiting customer of the queue (Figure 5).

Some examples of these types of systems are doctors' offices where there are several doctors, banks, box-offices, entrances to parking lots, parking spaces at bus terminals, schools and stores.

Given the assumption that there is no difference between the servers as they perform the same operation, the system's total capacity is in proportion to the number of parallel servers it has, in other words, if the system has c (c=1,2, …, n) servers, then the total capacity will be $c\mu$.

Figure 5. Queue with several servers

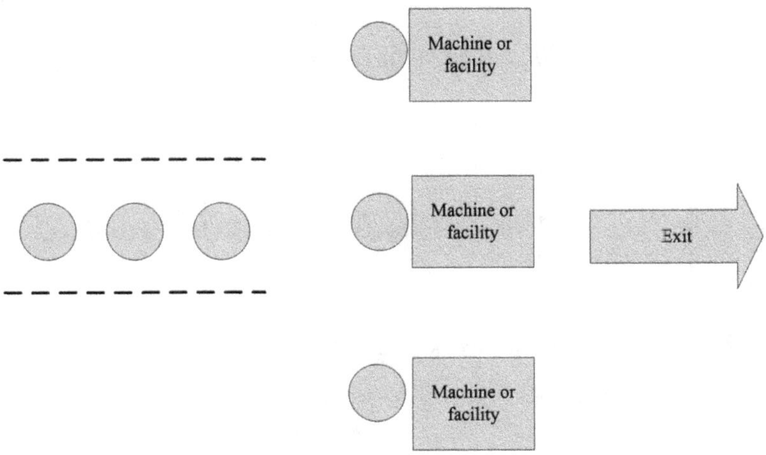

Series Systems

In this case the complete service requires several phases or stages. A factory has production lines consisting of a set of operations in a series. The production piece or order travels along the line going through one stage at a time and there is a transformation in each stage. The route is the same for each customer and assumes that entry is exclusively through the first stage or station of the process, the customer is not reworked or abandoned and is not allowed to leave until the stages are completed (Figure 6).

In restaurants, a customer generally arrives and puts their name on a list, then waits to be assigned a table. In fast food companies with the option to order from the car, the customer arrives at the place and stops at the first stage where they place their order, then go to the next stage where they pay and their order is delivered.

Systems With a Network Layout

Sometimes customers or pieces that are in the process follow a specific route within the process, because there are several options for paths to be followed. When these paths are represented, the system looks like a network with the stations interconnected. Likewise, those products that are nonconforming must be sent back for reworking or go through some stage for their correction before returning to their original route (Figure 7).

Figure 6. Queue system with stages arranged in series

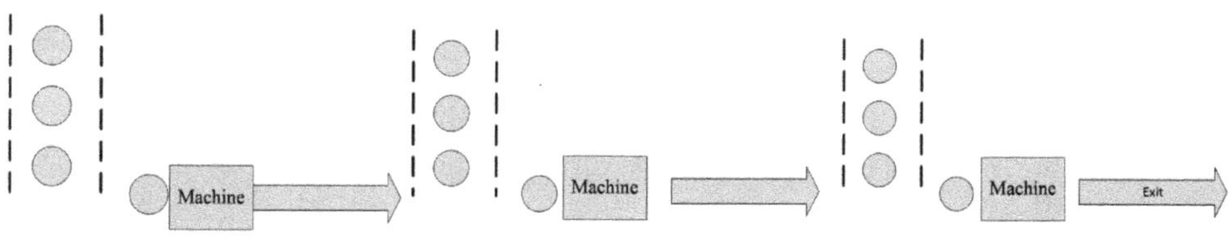

Figure 7. Queue system with network structure

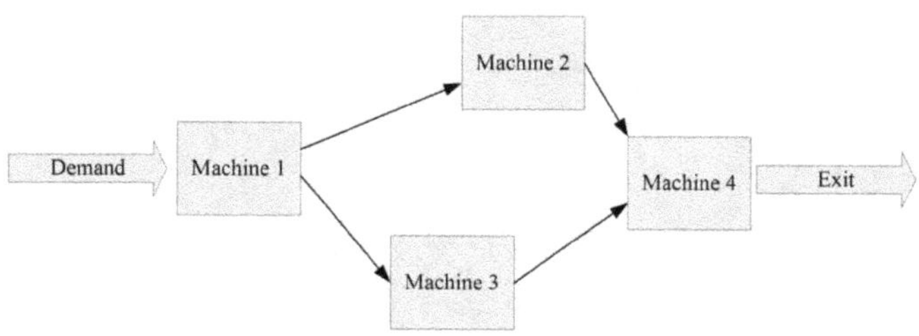

Network layouts is suitable for representing systems such as movie-theaters with their box offices, candy stalls, cafeterias, automated box offices and restroom areas; different departments of a hospital such as A&E, X rays, etc.; production systems with several lines; or a road infrastructure; production systems with their machines and the different products they produce.

EXAMPLES

Banks

In these types of systems, the customer arrives at the branch and waits in line to be served. There are a line of tellers in the customer service area where several customers are dealt with at the same time. The arrivals phenomenon is random. It is worth mentioning that there are days when there is greater demand, such as paydays or the end of month when people come in to make their payments. In these periods, managers usually assign more personnel to the tellers' area to speed up the flow of customers.

Parking Lots

The customers are the vehicles driven by people who want to park them. The service consists of two stages: the vehicle approaches the entrance to the parking lot and then drives around the parking lot looking for a space; when it finds a free space, the vehicle parks and stays at that stage for as long as the driver so requires. In the last stage, vehicle leaves the parking space (assuming they have already paid) and heads for the exit.

Mass Production Lines

Factory production usually consists of a series of steps or stages, where each station has a capacity to produce a certain quantity of material that is then sent on to the next stage. Once all the production stages have been covered, the material leaves as finished product. The demand exclusively enters in the first stage and the end product exclusively leaves through the last station.

A queue with material is formed in front of each station and this is called the product in process. One case that is more strictly in accord with reality is where there is only enough space to keep a limited quantity of product, in other words, the size of the queue has a maximum value constraint. This situation creates huge challenges for administrators as, limiting the amount of work in process means that the amount being produced is also limited. These types of systems shall be analyzed further on.

Flow of People Through Corridors

A system of corridors and stairways is similar to a queueing network with limited capacity. The models can be used to calculate the evacuation time and analyze the impact of adaptations (putting additional exits in, directing the flow through corridors), in order to improve the flow of people towards the exits.

In places like Mexico City there are rigorous regulations for the management of government buildings and offices. One example of a mandatory statute is the requirement for earthquake evacuation drills. Evacuation times are critical in this case, so a building needs to have enough exits for people to be able to evacuate it quickly.

Supply Chains

Nowadays managers and administrators see companies as entities, links or nodes in a connected network. A supply chain or network involves the suppliers of the raw materials, factories, means of transport, warehouses and end customers. The nodes in this network are connected to each other, at the very least. A production order does not immediately arrive at the warehouse as finished product; on the contrary, each order goes through every stage where it is gradually transformed until it reaches the customer as an end product. Administrators have to deal with situations arising from the random nature of demand, faults in the equipment, production capacity and the variety of products. These elements affect the time each production order stays in the chain (the residence time).

The residence time of an order is a performance measurement that enables us to estimate the response time as well as being a measurement of customer satisfaction. This system can be modeled as a queueing network. If a manager wants to make improvements in the response time, they may avail of an analysis that can identify the stages where the production has a slower flow (bottleneck) and estimate how much the flow can be improved by adding capacity or lessening the variability in process time on a specific production line.

Performance Measurements

A manager or administrator needs to have elements to hand that enable them to know how the system under their responsibility is performing. How do they know whether a bank has enough cashiers to meet the demand for service? How do they measure the time a customer remains waiting in the line? If a new cashier is added, how much will the investment cost?

One strategy for answering these questions would be to experiment directly with the system, for example by putting in cashiers and observing the result; however this is not always possible: If the system is a production line, top management can hardly give itself the luxury of interrupting the manufacture of a car to carry out some tests that would imply the consumption (and sometimes the waste) of resources. The situation is even worse if there is no system. How do we experiment?

Equations are a tool that allows us to understand how the various components of a system interrelate. Queueing systems are analyzed by means of equations that make it possible to have a set of performance measurements that can give us an idea about how they operate.

Table 2 shows the most common performance measurements. The basic performance measurements for queues are the cycle time (3) and (4), work in process (5) and (6) as well as the available capacity (7), and a set of equations have been developed that give information about the relationship between the elements of a queueing system and the impact they have on the time a customer stays in the system or on the number of pieces waiting for their turn to be processed (Table 2).

Cycle Time

This is divided in two: cycle time in the queue and cycle time in the system. The former is the time that the customer is in a queue waiting for service. The latter is the sum of the cycle time in the queue and the expected value of the service time.

The cycle time in the system is the time a customer is there from the moment the customer enters; lines up and goes through the service period until it leaves the system (Figure 8).

Table 2. Performance measures and equations

Name	Symbol	Units	Equation	
Average Cycle time in the system	CT_s	seconds, minutes, hours, days	$CT_S = CT_q + E\left[t_S\right]$	(3)
Average Cycle time in the queue	CT_q	seconds, minutes, hours, days	$CT_q = CT_S - E\left[t_S\right]$	(4)
Average Work in process in the system	WIP_s	Entities, pieces, clients	$WIP_S = \sum_{n=0}^{\infty} np_n$	(5)
Average Work in process in the queue	WIP_q	Entities, pieces, clients	$WIP_q = \sum_{n=n+1}^{\infty} \left(n-c\right)p_n$	(6)
Congestion	ρ	Dimensionless	$\frac{Demand}{Capacity} = \frac{\lambda}{c\mu} = \frac{\lambda t_S}{c}$	(7)

where:

- CT_S: Average cycle time in the system
- CT_q: Average cycle time in the queue
- WIP_S: Work in process in the system
- WIP_q: Work in process in the queue
- ρ: Congestion
- t_s: Service Time
- λ: Arrival rate of customers, demand
- μ: Service rate
- c: server's number
- p_n: probability of n customers in the system

Figure 8. Cycle time

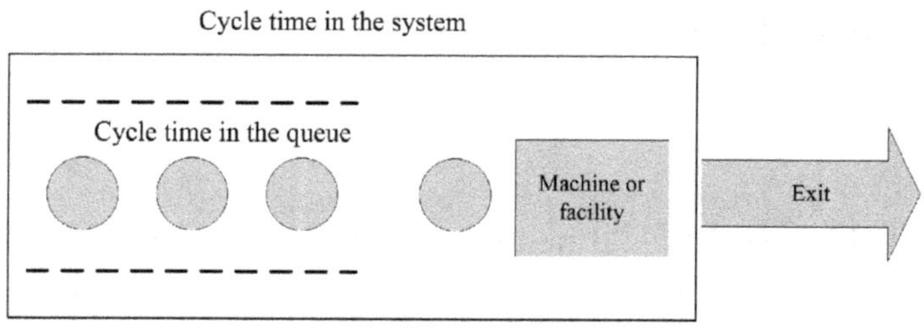

Figure 9. Work in process

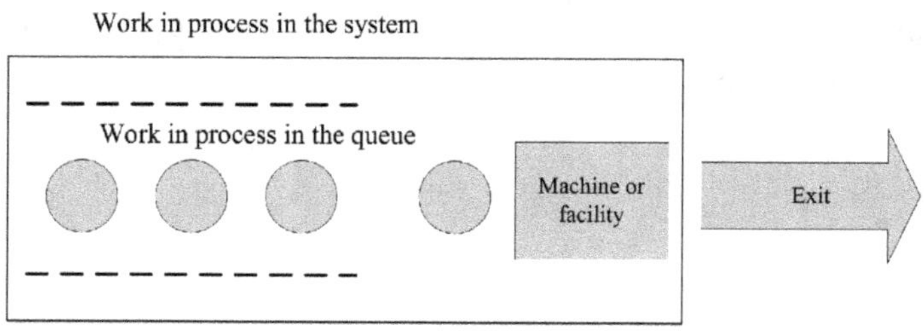

Work in Process

It is also necessary to point out that there is work in process in the queue and work in process in the system. As its name implies, the former refers to customers that are exclusively lined up waiting for service (WIP_q); while the latter (WIP_S) is the sum of the ones lined up and the ones being served (Figure 9).

Congestion

This is the measurement that gives the proportion of capacity being used for a particular demand for service. This is also known as traffic intensity and is represented by the Greek letter ρ (rho). The congestion value must be less than 1 for the system to be considered stable and if not, that is a sign that the system does not have enough capacity to meet the demand for service and consequently the line or queue will grow unceasingly. This situation is commonly to be found, for example, in box-offices when there is a much awaited premiere or in public hospitals where the numbers of patients are constantly increasing.

Little's Law

This is an equation that was developed and tested by D.C. Little in 1961 and links the amount of work in process to demand and cycle time. The equation is as follows:

$$CT = \frac{WIP}{\lambda} \tag{8}$$

Equation (8) indicates that the cycle time, CT, is obtained by dividing the amount of work in process, WIP, by the demand λ (customers arriving per unit of time). As we can see, when the demand for a service and one property (cycle time or work in process) are known, we can calculate the rest; for example, if the average amount of work in process in the system has already been calculated, then the average cycle time for a customer or piece in the system can also be calculated.

Let us consider a time interval (a, b). Now let us assume that the demand for service in this period is N_{ab} customers. By identifying each customer that arrives in the system from customer 1, 2, up to customer *N*, we find that the residence time for each customer in the system during this time period will be (Curry and Feldman, 2009) (Figure 10):

$$TC(a,b) = \frac{1}{N_{ab}} \sum_{i=1}^{N_{ab}} \left(\text{Departure Time}_i - \text{Arrival Time}_i\right) = \frac{1}{N_{ab}} \sum_{i=1}^{N_{ab}} \left(t_i^d - t_i^a\right) \quad (9)$$

Area *AB* between the arrivals curves $A(t)$ and curve for the customers leaving $D(t)$ for the interval $a < t < b$, is represented by the sum total. Equation (9) is the area between the curves and can be represented with the following integral:

$$AB = \int_a^b \left(A(t) - D(t)\right) dt \quad (10)$$

The integral (10) represents the number of customers in the system because the difference $N(t) = A(t) - D(t)$ is the number of customers in the system

Figure 10. Arrival curve A(t) and Departure curve D(t)

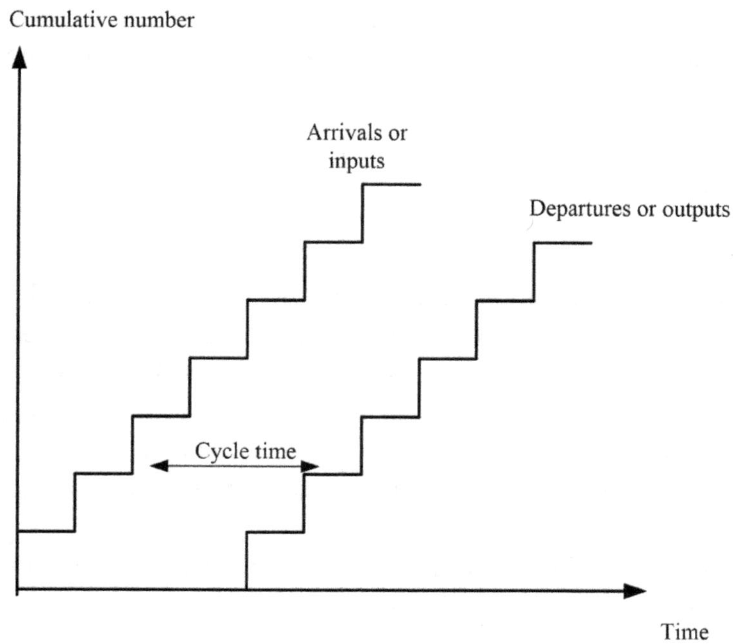

in instant *t*. The average number of customers is obtained by dividing it by the total time:

$$WIP(a,b) = \frac{1}{b-a} \int_a^b \left(A(t) - D(t)\right) dt \tag{11}$$

If the area between curves $A(t)$ and $D(t)$ in (11) is a constant, then the average number of customers in the system is:

$$WIP(a,b) = \frac{1}{b-a} AB \tag{12}$$

The cycle time is the average number of customers in instant *t* divided by the number of customers that arrive during the period (a, b):

$$CT(a,b) = \frac{1}{N_{ab}} AB \tag{13}$$

Starting from (12) and (13) and remembering that the number of customers in instant *t* is $N(t)$, the work in process will be:

$$WIP(a,b) = \frac{N}{b-a} CT(a,b) \tag{14}$$

The demand for service (λ) is the number of customers that arrive during a period, that is to say: $\frac{N_{ab}}{b-a}$, substituting in (14), we have:

$$WIP(a,b) = \frac{N_{ab}}{b-a} CT(a,b) \tag{15}$$

The term $\frac{N_{ab}}{b-a}$ in (15) is the number of customers that arrive during a period of time (λ), and we shall also refer to it either as the demand for service or simply as the demand:

$$WIP(a,b) = \frac{N_{ab}}{b-a} CT(a,b) \qquad (16)$$
$$WIP = \lambda CT$$

Little's Law (16) provides the average long-term value for both the work in process and the cycle time. A long-term operation implies that the mean and standard deviation do not vary over time, in other words, they reach a limit value. It is worth mentioning that Little's Law gives parameters that are measurable in any system and is indifferent to the type of situation, in other words, the equation applies equally to the field of manufacturing or the field of services.

In essence, Little's Law states that the number of customers that enter the system through one end is equal to the number of customers that leave it through the other end, and they will stay for a period that is called the cycle time. As demand rises, in other words, if more customers arrive and there are no changes in the system's capacity, the work in process will increase and the average cycle time will also increase, in consequence.

Kendall's Notation

As the reader will have already observed, there are a number of elements we need to take into account when we identify a queue: number of servers, characteristics of the demand for service, characteristics of the service, space available in the system, to mention just a few. In 1953, David George Kendall designed a notation that allows us to classify a queue according to the characteristics of the time that elapses between consecutive customers arriving in the system, the characteristics of the station's service time and the number of servers. This notation has been amplified to include the service discipline and the number of spaces available for the customers in the system.

Arrivals/Service/Number of servers/Discipline/Capacity

The first symbol corresponds to the probability distribution for the time that elapses between two consecutive customers. The second symbol is the probability distribution for the service time. The third symbol is an integer number for identifying the number of servers. The fourth symbol is for the rule or discipline for customer service. And finally the last symbol is used to define the number of places available for customers including the ones being served (Table 3).

Table 3. Symbols of Kendall notation

Element	Symbol
Arrives:	
Exponential Deterministic Erlang General	M (Markovian) D E G (Any pdf with mean t_a and variance σ^2)
Service Time:	
Exponential Deterministic Erlang General	M(Markovian) D E G (Any pdf with mean t_s and variance σ^2)
Number of servers	c
Discipline	FIFO, LIFO, P, SIRO
Capacity	
No limit Limited capacity	∞ K

EXAMPLES

Box-Offices

Let us assume that the customers arrive at a box-office with a certain periodicity which follows an exponential distribution and the service time that the vendor invests in the operation (welcoming them, asking which movie they want to see, receiving money, giving the change back and supplying the tickets) also follows an exponential distribution and there is a single box office in operation. The customers are served under the first in, first out discipline and usually there are no restrictions on the number of customers that can approach the box-office. In Kendall's notation, this has the following representation:

M / M / 1 / FIFO / ∞

Now let us assume that 2 additional tills are opened in the box-office to meet the demand of the customers. None of the other elements changes. In Kendall's notation, the system now has the following representation:

M / M / 3 / FIFO / ∞

A&E Department in a Hospital

Let's take the example of a hospital's A&E Department. Patients go to a section where there are 4 doctors and are served according to the level of emergency, in other words, patients with injuries or symptoms that indicate a risk or danger of loss of life have a high value or priority for attention and are seen immediately, whereas patients with minor symptoms or whose lives are not at risk are given a low value or priority for attention. Assuming that the arrivals follow an exponential distribution, the time the doctors take to attend a patient follows a general probability distribution with a mean service time t_s and standard deviation σ_{ts}. Finally there are no restrictions on receiving patients:

M / G / 4 / P / ∞

Factories

A machine in a factory that processes one piece at a time; with a Poisson type demand; and an exponential distribution of service time. The space available for lining up the pieces to be processed is 4 places.

M / M / 1 / FIFO / 5

Exponential Distribution

A non-negative variable *T* follows an exponential distribution with parameter λ, if its density function is:

$$f_T\left(t\right) = \lambda e^{-\lambda t} \quad t \geq 0 \tag{17}$$

Starting from (17), we get the probability function (18):

$$F_T\left(t\right) = 1 - e^{-\lambda t} \quad t \geq 0 \tag{18}$$

The expected value and the variance of a distributed random variable with the exponential function are, respectively, $E\left[T\right] = \frac{1}{\lambda}$ and $Var\left[T\right] = \frac{1}{\lambda^2}$. For the analysis of stochastic processes in general and queues in particular, exponential distribution has a very important property, which is the absence of memory. Let us assume that variable T is the lifetime of a piece and that t is the time that the piece has survived without failing, what will the conditional probability of the piece that is still working in $T - t$ time be?

First of all, the probability for a variable $x > 0$ of the piece still working after a time $T-t$ is expressed as a conditional probability $\Pr\left\{T - t > x \mid T > t\right\}$:

$$\Pr\left\{T - t > x \mid T > t\right\} = \frac{\Pr\left\{T > t + x, T > t\right\}}{\Pr\left\{T > t\right\}} \tag{19}$$

If the variable follows a probability exponential distribution then:

$$\Pr\left\{T - t > x \mid T > t\right\} = \frac{\Pr\left\{T > t + x\right\}}{\Pr\left\{T > t\right\}} = \frac{e^{-\lambda(t+x)}}{e^{-\lambda t}} \tag{20}$$

From expression (20):

$$\Pr\left\{T - t > x \mid T > t\right\} = e^{-\lambda t} \tag{21}$$

The result (21) establishes that the conditional probability is

$$\Pr\left\{T - t > x \mid T > t\right\} = e^{-\lambda t},$$

in other words, the probability of a piece still working after instant t is independent of the time it has been working.

EXERCISES

1. In the following systems, construct a drawing or diagram and identify: the server, the customers, the number of stages, and the income of demand.
    a. A parking lot
    b. Clinical analyses
    c. A candy stall in a movie-theater

   d. Driver's license office
   e. Bus station
   f. Airport
   g. Restaurant
   h. Drive-thru service at a fast food restaurant
2. Use Kendall's notation for the following systems:
   a. Demand with exponential times, Exponential service, 4 servers, First In, First Out discipline
   b. Demand with normal inter-arrival time distribution, Exponential service, 1 server, Last In, Last Out discipline
   c. Demand with normal inter-arrival time distribution, Exponential service, 1 server, Last In, Last Out discipline
   d. Demand with exponential inter-arrival time distribution, exponentially distributed service, 1 server, first in, first out discipline, system capacity of 5 units (including the server)

REFERENCES

Curry, G. L., & Feldman, R. M. (2011). *Manufacturing systems modeling and analysis* (2nd ed.). Berlin: Springer. doi:10.1007/978-3-642-16618-1

Gross, D., Shortle, J. F., & Thompson, J. M. (2008). *Fundamentals of queueing theory*. Hoboken, NJ: J. Wiley & Sons. doi:10.1002/9781118625651

Hall, R. W. (1991). *Queueing methods: For services and manufacturing*. Englewood Cliffs, NJ: Prentice Hall.

Render, B., & Stair, R. M. (1997). *Quantitative analysis for management*. Upper Saddle River, NJ: Prentice Hall.

Taylor, H. M., & Karlin, S. (1984). *An introduction to stochastic modeling* (3rd ed.). Orlando, FL: Academic Press.

Chapter 2
Review of Probability Distributions

ABSTRACT

In Chapter 2, probability distributions are presented; the distributions exposed are those with more relation to the analysis and study of waiting lines; discrete distributions: binomial, geometric, Poisson; continuous distributions: uniform, exponential, erlang, and normal. Confidence intervals are calculated for some of the parameters of the distributions. A brief example of the generation of pseudorandom exponential times using a spreadsheet is presented. The chapter closes with the goodness-of-fit tests of probability distributions, especially the Anderson-Darling test. The statistical language of programming R is used in the exercises performed. Several codes are proposed in R Language to perform calculations automatically.

PROBABILITY DISTRIBUTIONS

When an experiment with random results is made. The set of possible events is called Sample Space. A function or mapping that identifies the events of the Sample Space with the set of real numbers is called Random Variable.

In general terms, a probability distribution could be called a rule for assigning probability values to random variables. Random variables can be classified in two types: discrete and continuous. Continuous variables are real numbers represented by X, while categorical variables are integers represented by N. Integers are a subset of real numbers.

DOI: 10.4018/978-1-5225-5264-2.ch002

- **Cumulative Distribution Function:** One particularly useful way of expressing probability distribution is by means of mathematical expressions: one is called probability density function, represented by f(x). The other type is called cumulative distribution function (CDF), represented by $F_X(x)$ or F(x), and assigns probabilities to intervals of the random variable.

A cumulative distribution function can be expressed as:

$$P(X \leq x) = F_X(x) = \int_{-\infty}^{x} f(x)\,dx \text{ X continuous}$$
$$P(N \leq k) = F_N(k) = \sum_{k=-\infty}^{k} f(k) \text{N discrete} \tag{1}$$

where: *F(x)* is the cumulative distribution function; *f(x)* is the probability density function; X is a continuous random variable; "x" (lower case x) is a numeric value of interest. N is a discrete random variable and finally K is a continuous numeric value of interest.

Probability functions contain parameters that give every distribution different properties and behavior. These parameters are classify into three types: location parameters, scale parameters and shape parameters.

Every distribution has a different combination of parameters that, together with the functional form of the mathematical expression, gives different properties to every probability function.

Additional functions have been defined that help us to understand the behavior of a variable random. These functions are particular to each random variable. The ones most used in queues, are: expected value, variance and coefficient of variation:

- **Expected Value:** Expected value gives the value where the "center of gravity" of the probability distribution is. This is the point of equilibrium for distribution and is expressed as

$$E[X] = \int_{-\infty}^{\infty} xf(x)\,dx \text{ X continuous}$$
$$E[N] = \sum_{k=-\infty}^{\infty} kf(k) \text{N discrete} \tag{2}$$

- **Variance:** Variance is a measure of dispersion of the random variable distribution and is defined as:

$$V[X] = \int_{-\infty}^{\infty} (x - E[X])^2 f(x)\,dx \text{ X continuous}$$
$$V[N] = \sum_{k=-\infty}^{\infty} (k - E[N])^2 f(k) \text{N discrete} \tag{3}$$

- **Coefficient of Variation:** The coefficient of variation is the division by the square root of the variance, called the standard deviation, and the expected value of the random variable is equally defined as a discrete or a continuous variable:

$$C(X) = \frac{\sqrt{V[X]}}{E[X]} \tag{4}$$

It is often more practical to use the coefficient of variation squared, which in this case is:

$$C^2[X] = \frac{V[X]}{E^2[X]} \tag{5}$$

- **Quantile:** Assuming that we have a cumulative probability value, p = F(x), for a numeric value x, the value of x for the random variable that corresponds to that cumulative probability is called quantile (p). A quantile) or q(p), complies with:

$$p = F_X(quantile(p)) = \int_{-\infty}^{quantile(p)} f(x)\,dx \text{ X continuous} \tag{6}$$

If p=F(x), then $q(p)= F^{-1}(p)= x$.
Given a sample of *n* randomly observed data, $x_1, x_2, \ldots, x_n$

- They are ordered in a non-descending order $x_{(1)} \leq x_{(2)} \leq \ldots \leq x_{(n)}$, where index (i) indicates the order of the datum.

- If we wish to calculate the quantile corresponding to probability p, we get: (j)= $n*p$.
- If (j) is an integer, the quantile is estimated as $=(x_{(j)}+x_{(j+1)})/2$; if (j) is not an integer, the integer part of (j), represented by [j], is taken exclusively and the quantile shall be: $x_{([j]+1)}$

For example: we have a sample of 20 data that have already been ordered as shown in Table 1.

If we want to get the quantile corresponding to a p value=0.5, also called a median, (j)=20*0.5=10 is calculated. This is an integer. Then the quantile for 0.5 shall be =(x(10)+x(10+1))/2= (10+10)/2= 10. The median of these data is 10.

If we want to calculate the quantile for p=0.632, we get (j)=0.632*20=12.64, which turns out not to be an integer, then the quantile will be $x_{([12.64]+1)}=x_{(12+1)}=x_{(13)}=11$. In other words, 63.2% of the data is estimated as being less than 11.

Uniform Distribution

A uniform distribution with a simple structure for continuous variables that is very useful for simulating random events. It is defined as the probability function that assigns uniform probability values to a continuous random variable X, within a [a, b] interval:

$$f\left(x\right)=\frac{1}{b-a}a\leq x\leq b \tag{7}$$

where: f(x) is the probability density function; "a" is the lower bound of the interval for X, while "b" is the upper bound.

The corresponding cumulative distribution function is:

Table 1.

(i)	1	2	3	4	5	6	7	8	9	10	11	12	13	14	15	16	17	18	19	20
x	4	4	5	6	7	7	8	9	10	10	10	11	11	14	18	19	20	24	28	38

$$F\left(x\right)=P\left(X\leq x\right)=\frac{x-a}{b-a}a\leq x\leq b \tag{8}$$

where: F(x) is the cumulative distribution function; "x" the numeric value for which the probability is calculated, "a" and "b" are respectively the upper and lower bounds for the variable.

The expected value, variance and coefficient of variation are respectively:

$$E\left[X\right]=\frac{a+b}{2};V\left[X\right]=\frac{\left(b-a\right)^{2}}{12};C\left[X\right]=\frac{b-a}{\sqrt{3}\left(a+b\right)} \tag{9}$$

One special case of this distribution is the uniform distribution for x in the interval [0,1], that gives us:

f(x)=1, F(x)=x, E[X]=0.5,V[X]=1/12,y C[X]=$1/3^{½}$.

- **Software Use:** Uniform pseudorandom data in the interval [0,1] can be generated on a computer and used to generate data in other probability distributions. One method of generating them is a multiplication method that generates a number with a lot of digits and then cuts some of them off to generate a new number that is again multiplied to generate another very long one that is, in turn, truncated to produce a new one.

For example, the following formulas are used on a spreadsheet:

- In Cell A1: we enter an initial number called the SEED.
- In Cell A2 = MOD(48271*A1, 2^31-1)
- In Cell B2 = A2/(2^31-1)

and this is copied to the following cells, using, for example, an initial seed value =123, which gives us the data shown in Table 2.

We can get pseudorandom data with a uniform distribution between 0 and 1, on spreadsheets by using the command RAND(). In the R statistical programming language, we can, if needed, get pseudorandom data with a uniform distribution using the command:

Table 2. Data generation using the congruential method, column B uniform contains data between 0,1

	A	B
1	123	
2	5937333	0.002765
3	9.86E+08	0.458991
4	2.08E+09	0.96637

runif(sample size, a,b)

Where this command generates a set of data in the interval [a,b] in the quantity marked as sample size.

Some Discrete Probability Distributions

There is a group of discrete probability distributions for random variables that are related. s is due to the fact that their probability assignment formulas depend on certain common assumptions that are important when phenomena, such as the arrival of customers, have to be analyzed.

1. **Bernoulli Trials:** We shall now proceed to define discrete random variables for the so-called "Bernoulli Trials". These trials generate discrete random variables in a series of k random experiments in such a way that:
2. Each experiment has only two possible results, identified as 0 and 1. These are also identified as: failure and success, arrival and non-arrival.
3. One of the two results is of interest and probability of it occurring is called "p". The value "p" of the result of interest is constant in all the experiments. The probability "p" is unaffected by knowledge of the previous results.
4. Sometimes the probability of the event that is complementary to the event of interest is designated as $q=1-p$.

Binomial Distribution

Repetitions of the Bernoulli trials is a binomial experiment. The binomial experiment is a sequence of Bernoulli trials. If several Bernoulli Trials are performed, the random variable N can be the number of times that a result of interest occurred. The distribution that assigns probabilities to this random variable is called binomial distribution, as it uses a binomial coefficient. The binomial density function for N, using p=λ/n, is:

$$f(k) = \frac{n!}{k!(n-k)!}(p)^k (1-p)^{n-k} = \frac{n!}{k!(n-k)!}\left(\frac{\lambda}{n}\right)^k \left(1-\frac{\lambda}{n}\right)^{n-k} \text{ for } k = 0,1,\cdots. \tag{10}$$

where: "n" is the number of trials done; "k" is the number of events of interest that are observed in "n" trials; "λ/n" the constant probability "p" of the event of interest in each one of the trials; while "λ" corresponds to the expected value of the random variable.

The expected value, variance and coefficient of variation are respectively:

$$E[N] = np = \lambda; V[N] = np(1-p) = npq; C^2[N] = \frac{1-p}{np} = \frac{q}{np} \tag{11}$$

where: p is the constant probability of the event of interest in n Bernoulli Trials; $0<p<1$.

- **Software Use:** Probability functions can be calculated using software, such as the R statistical language. In R for the binomial distribution with parameters "n" and "p" for the density function f(k), we have the function dbinom(k, n, p).

For the cumulative distribution function $F(k)$ we have pbinom(k, n, p).

The inverse cumulative distribution function, $F^{-1}(prob)$ is obtained using qbinom(probability,n,p).

For example:

For parameters: n=3, p=0.5, we have

```
f(1)=P(1)=
> dbinom(1,3,0.5)
[1] 0.375
f(0)=P(0)=
> dbinom(0,3,0.5)
[1] 0.125
F(1) = P(N≤1)
> pbinom(1,3,0.5)
```

```
[1] 0.5
F-1(0.5)=
> qbinom(0.5,3,0.5)
[1] 1
```

- **Simulation for a Binomial Variable:** The simulation of a binomial random variable on a computer can be done by counting the quantity of pseudorandom numbers (0-1) in an "n" sequence that are less than the parameter "p". For example, on a spreadsheet for n=10 we will get pseudorandom numbers (0-1) and a value of p=0.7, as shown in Table 3.

In the statistical programming languages R. the following command is used:

```
rbinom(sample size, n, p)
```

This command gives a sample size for data that have a binomial distribution with parameters "n" and "p".

For example: In order to generate 30 data that have a binomial distribution with parameters n=10 and p=0.7, we get:

```
> rbinom(30,10,0.7)
[1] 6 6 5 9 7 8 8 6 8 8 9 7 9 8 7 8 6 7 6 9 7 6 8 7 7 8 7 7 9 8
```

Table 3. Count of values less than 0.7 for generating a datum of one a binomial distribution with n=10 and p=0.7

	A	B
1	0.8951240871	=RAND()
2	0.078066197	=RAND()
3	0.3081485776	=RAND()
4	0.1583383474	=RAND()
5	0.345822661	=RAND()
6	0.7806633327	=RAND()
7	0.6983628406	=RAND()
8	0.8253552434	=RAND()
9	0.9677598099	=RAND()
10	0.8094264171	=RAND()
	5	=COUNTIF(A1:A10;"<0.7")

- **Estimator of "P":** To estimate the value of parameter "p" of the binomial distribution, the proportion is used from a sample of "n" observations with k results for the event of interest, which is calculated using the quotient "k/n".

For example: If k=221 results are observed for the event of interest in a sample of size n=300 observations, the estimator of the proportion is: 221/300 = 0.7366667.

Geometric Distribution

If a series of Bernoulli Trials are performed until the occurrence of the event of interest, which has a constant probability equal to "*p*" in every trial, and the number of trials performed until a result of interest is achieved is counted (in other words, the random variable *N* is the number of trials performed to achieve the first result of interest), then the probability distribution for this variable is:

$$\begin{aligned} f(k) &= p(1-p)^{k-1} = pq^{k-1} \\ F(k) &= 1-(1-p)^{k} = 1-q^{k} \end{aligned} \quad \text{for } k = 1,2 \tag{12}$$

where: *p* is the constant probability in all the trials for the event of interest. "*k*" is the number of trials we have to do to achieve the result of interest. "q" is the complement of "p", $q=1-p$.

The expected value, the variance and the squared coefficient of variation result in:

$$E[N] = \frac{1}{p}; V[N] = \frac{1-p}{p^2}; C^2[N] = 1-p \tag{13}$$

This distribution can be applied to a queuing phenomenon, assuming, for example, that only 1 or no customers can arrive for very small units of time. The probability of the arrival of a customer in that small time interval is *p* and the probability of no customer whatsoever arriving is $q=1-p$. Then the time T until the arrival of the next customer shall follow a geometric probability distribution.

If this is the case, what is the probability of no customers arriving in the next k units of time, P(T > k)?, Then:

$$P\left(T > k\right) = 1 - P\left(T \leq k\right) = 1 - F\left(k\right) = 1 - \left(1 - q^{k}\right) = q^{k} \tag{14}$$

This result can be explained as the calculation of the probability of not having arrivals in a sequence of k consecutive units of time, bearing in mind that the probability of arrival (event of interest) is p.

- **Lack of Memory:** Geometric distribution has a property that is illustrated by this case. Assuming that the event of interest is now a customer already inside the system leaving the service and knowing that a customer already has t units of time inside the service, what is the probability of there still being s additional units of time? This is posed as:

$$P\left(T > t + s \middle| T > t\right) = \frac{P(T > t + s)}{P(T > t)} = \frac{q^{t+s}}{q^{t}} = q^{s} = P(T > s) \tag{15}$$

It turns out that the conditional probability of an event of interest being observed in s instants of time, knowing that t has already past, is independent of the time t that has past.

This property is called "lack of memory" and is shared with another probability distribution for continuous variables which is called the exponential probability distribution.

- **Software Use:** For the probability calculations in the R statistical language for a random variable N with geometric distribution with parameter p, we have the following formulas:

For probability density function $f(k)$ we use dgeom(k-1, p).

For cumulative distribution function $F(k)= P(N\leq k)$ we use the command pgeom(k-1, p).

For example, if we have a geometric distribution with parameter $p=0.5$, for $k=1$ and k=2 we get:

$f(2) =P(2)=0.5(1-0.5)^{1}$

```
> dgeom(1,0.5)
[1] 0.25
f(1) =P(1)= 0.5(1-0.5)⁰
> dgeom(0,0.5)
[1] 0.5
F(1)=P(N≤1)=P(1)= 1-0.5¹
> pgeom(0,0.5)
[1] 0.5
F(2)=P(N≤2)=P(1)+ P(2)=1-0.5²
> pgeom(1,0.5)
[1] 0.75
```

In these R language formulas we use the value *k-1* rather than directly using the value k. In R, the geometric random variable was defined as the number of distinct events before the first event of interest. However, here the random variable was defined as the number of trials required for the first event of interest to occur.

- **Simulation of Geometric Variables:** In order to simulate data with a geometric probability distribution, Data can be generated on a spreadsheet using the following formula:

$$k = \frac{\log\left(rand\right)}{\log\left(1-p\right)} \tag{16}$$

where: *rand* is a pseudorandom number of uniform distribution in the interval (0,1); "p" is the geometric distribution parameter; "k" is the smallest integer that is higher than or equal to log(rand)/log*(1-p)*. This is obtained on a spreadsheet using the function CEILING(). The use of the commands in a spreadsheet is as shown in Table 4.

In the R statistical programming language, the generation of geometrically distributed data is achieved by using the command: rgeom(simple size, p) +1,

A unit has to be added to the value calculated by R, as a geometric variable is defined in R as the number of attempts before getting the first result of interest. Here the random variable counts the number of attempts to get the first result of interest, so it will always be a larger unit. For example, for a sample size of 30 and a value of p =0.7, it calculates:

```
> rgeom(30,.7)+1
[1] 1 1 2 1 1 1 3 1 1 1 1 4 1 1 1 2 1 3 2 6 1 2 1 1 1 1 1 1 1 1
```

Table 4. Generation of 10 data with geometric distribution, with value p =0.7

	A	B
1	1	=CEILING(LN(RAND())/LN(1-0.7),1)
2	4	=CEILING(LN(RAND())/LN(1-0.7),1)
3	1	=CEILING(LN(RAND())/LN(1-0.7),1)
4	4	=CEILING(LN(RAND())/LN(1-0.7),1)
5	3	=CEILING(LN(RAND())/LN(1-0.7),1)
6	2	=CEILING(LN(RAND())/LN(1-0.7),1)
7	1	=CEILING(LN(RAND())/LN(1-0.7),1)
8	1	=CEILING(LN(RAND())/LN(1-0.7),1)
9	1	=CEILING(LN(RAND())/LN(1-0.7),1)
10	1	=CEILING(LN(RAND())/LN(1-0.7),1)

- **Estimation of p:** We can estimate the value of p from a sample of n observed data by using the inverse of the arithmetic average of the observed sample. Thus, for example, if the sample gives an average of 1.22, the estimated value of p will be its inverse: 0.8196721.

Poisson Distribution

Continuing with the Bernoulli Trials, the binomial distribution calculates the probability of getting "k" events of interest for "n" trials done.

Now if we examine the events in an interval of time, assuming that the period of time is divided into *n* very small intervals of the same size, the intervals are considered to be so small that the probability of the occurrence "p" of the event of interest is also small.

The time interval is assumed to be so small that it can only be observed in a Bernoulli Trial. Then the event of interest may or may not occur in each time interval.

Moreover, if we observe the above situation for a sufficiently long period of time, we will get a set of long-term observations.

An average of the observed events of interest can be estimated in that long-term period. The average can be called λ (lower case lambda). Lambda may be for example: 5 arrivals of customers per hour, 28 sales a day. Therefore the average per unit of time λ can be equal to the average of a very long sequence of "n" Bernoulli Trials, with every trial having a "p" probability of occurrence $\lambda = np$. Then we would estimate that constant p is:

$$\lambda = np \tag{17}$$

If N is defined as a random variable that counts the number of events of interest in a time interval. We get a density function, for a value "n" that tends to grow a lot and an average constant value λ:

$$f(k) = \lim_{n\to\infty}\left[\frac{n!}{k!(n-k)!}\left(\frac{\lambda}{n}\right)^k\left(1-\frac{\lambda}{n}\right)^{n-k}\right] = \frac{\lambda^k e^{-\lambda}}{k!} \text{ for } k = 0,1,\cdots. \tag{18}$$

where: The function between the square brackets is the binomial probability function. The number "e" is the so-called natural number. "e" is the basis for natural logarithms, approximately 2.718281828459.

The resulting probability density function is called Poisson distribution. The expected value, variance and coefficient of variation of the Poisson distribution are:

$$E[k] = \lambda \tag{19}$$

$$V[k] = \lambda \tag{20}$$

$$C[k] = \frac{1}{\sqrt{\lambda}} \tag{21}$$

- **Software Use:** To calculate the probabilities of a random variable with Poisson probability distribution using with a parameter λ.

To calculate the density function *f(k),* we use the command dpois(k, λ)

For cumulative distribution function *F(k)= P(N≤k)* we use the command ppois(k, λ).

- **For Example:** For a Poisson random variable with λ (lower case lambda) equal to 6, for the vector of values k= (0, 1, 2, 3, 4, 5, 6):

> k<-c(0,1,2,3,4,5,6) #here vector k is defined with the instruction "c", concatenate or combine.

```
> dpois(k,6)
[1] 0.002478752 0.014872513 0.044617539 0.089235078 0.133852618
0.160623141
```

[7] 0.160623141 #the [1] and [7] respectively indicate the first and seventh elements of the response vector.

```
> ppois(k,6)
[1] 0.002478752 0.017351265 0.061968804 0.151203883 0.285056500
0.445679641
```

[7] 0.606302782 #the [1] and [7] respectively indicate the first and seventh elements of the response vector.

- **Another Example:** For a binomial distribution with p=0.0003 and n=10,000, the probability of observing more than one result is calculated with $P(N>1)=1-P(N\leq 1) = 1- F(1)$. Then to calculate F(1) we use the command for cumulative distribution of the binomial: pbinom(1,10000,0.0003). It can also be approximated with the Poisson distribution using $\lambda= np= 10{,}000*0.0003$. In which case F(1) is calculated using the corresponding command ppois(1,3. Then we have:

```
> 1-pbinom(1,10000,0.0003)
[1] 0.8008965
> 1-ppois(1,3)
[1] 0.8008517
```

- **Poisson Random Data:** To generate random data with a Poisson distribution in the R statistical programming language we use the command rpois(sample size, λ). The rpois function generates a data sample with a Poisson probability distribution and parameter λ.

For example, for a sample size of 20 and a value of λ=5, we get:

```
> rpois(20,5)
[1] 4 3 5 4 6 5 5 4 8 4 4 6 4 5 3 5 8 1 3 3
```

where: the symbol “>” means the start of a command line. [1] indicates the first element of the response vector.

- **Example of Estimation of λ:** To estimate the parameter λ, which corresponds to the mean of the Poisson probability distribution based on a sample of observed data, we can use the arithmetic average.

For example, the emission of alpha radioactive particles was found to follow a Poisson distribution (Rutherford, E. Geiger, H. & Bateman, H., 1910). During 2,608 periods of time, the emission of alpha particles was counted and the data shown in Table 5 was obtained.

This gives a total of

n=2,608 periods = 57+203+…+4+1

and a total,

K events of interest = 57*0+203*1+383*2+525*3+…+1*14=10,097

with the average per period being: 10,097/2608 = 3.87.

Therefore the estimation of lambda is 3.87.

- **Confidence Interval:** If we want a confidence interval for the estimation of lambda we use the chi-squared distribution (χ^2).

with the lower bound for lambda being calculated with a probability of α/2. The upper bound of the interval is calculated for a probability of 1-α/2. Thus a region of probability of 1-α is constructed between the two bounds. The bounds are calculated:

Table 5.

Number of Particles (k)	0	1	2	3	4	5	6	7	8	9	10	11	12	13	14
Frequency	57	203	383	525	532	408	273	139	45	27	10	4	1	4	1

$$\lambda_L = \frac{1}{2n}{}^{2}_{2K,\alpha/2}$$
$$\lambda_U = \frac{1}{2n}{}^{2}_{2K+2,1-\alpha/2} \quad (22)$$

where: λ_L is the calculated lower bound. "n" is the total number of observed frequencies, K is the total of events of interest observed. 1-α is the desired confidence level. The parameter ${}^{2}_{df,prob}$ is the value of a chi-square for "df" degrees of freedom and a cumulative probability "prob". If the degrees of freedom turn out to be zero, the value of the chi-square is taken as zero. λ_U corresponds to the upper bound calculated for λ.

The chi-square values can be calculated in the R statistical programming language using the command: qchisq(probability, degrees of freedom).

For the values in the above example with a α=0.05, 1-α=0.95, we get:

```
> frec<-c(57,203,383,525,532,408,273,139,45,27,10,4,0,1,1)
> n<-sum(frec); n
[1] 2608
> k<-0:14; k
[1] 0 1 2 3 4 5 6 7 8 9 10 11 12 13 14
> K<-sum(frec*k); K
[1] 10097
> lambda<-K/n; lambda
[1] 3.871549
> lambda_L<-qchisq(0.025,2*K)/(2*n); lambda_L
[1] 3.796397
> lambda_U<-qchisq(0.975,2*K+2)/(2*n); lambda_U
[1] 3.947814
```

It turns out that the average is estimated as 3.871549 and has a 95% confidence interval of (3.796397, 3.947814). This is always provided the Poisson distribution model is the proper one. This has been proven by the authors.

- **Chi-Square Goodness-Of-Fit Test:** The chi-square goodness-of-fit test is a test for assessing the goodness of fit in the event that the data appear in the form of class marks and their frequencies, such as when there is a histogram.

The observed frequencies are compared with the expected frequencies. The expected frequencies are estimated using the distribution we wish to test. Then a chi-square statistic is calculated using the following formula:

X2<-sum((O-E)^2/E

where: O is the observed frequencies vector and E is the expected frequencies vector. The calculated chi-square statistic has degrees of freedom equal to the number of classes considered less the number of parameters estimated for distribution minus one.

If the statistic is in the rejection region or the P value associated with the statistic is higher than the alpha value being considered, the null hypothesis that the data are distributed according to the assumed function is rejected.

Continuing with the example, the observed frequencies have to regroup the last values as the chi-squared test needs the number to be compared for any class to be higher than 5. Thus the Observations vector "O" is:

```
> O<-c(57,203,383,525,532,408,273,139,45,27,16); O
[1] 57 203 383 525 532 408 273 139 45 27 16
```

The classes are then from 0 to 9 and one to the end for all the values higher than or equal to 10 with a frequency of 16. Now the "E" expected frequencies are calculated assuming that the probability distribution is a Poisson with a rate of 3.947814. Then the probabilities obtained are multiplied by the total of the observations and the class of the last one is adjusted for difference to made the total sum of frequencies for "O" Observed and "E" Expected frequencies fit:

```
> k<-0:9; k
[1] 0 1 2 3 4 5 6 7 8 9
> E<-dpois(k, 3.871549)*2608; E #n = 2608
[1] 54.31443 210.28097 407.05655 525.31312 508.44387 393.69307
254.03367
[8] 140.50054 67.99434 29.24927
> E[11]<-2608-sum(E);E # the final class is adjusted for
difference
[1] 54.31443 210.28097 407.05655 525.31312 508.44387 393.69307
254.03367
[8] 140.50054 67.99434 29.24927 17.12015
```

The chi-square statistic is calculated using the sum of the squared difference between observed frequencies minus expected frequencies and dividing each difference by the corresponding expected value:

```
> X2<-sum((OR-E)^2/E); X2
[1] 12.87262
```

To get the P value we use the command pchisq with the value of the resulting statistic "X2" and the degrees of freedom = 11-1-1=9, 11 classes minus 1 parameter minus 1:

```
> pchisq(X2,9)
[1] 0.8315445
```

The P value is higher than the usual alpha values: 0.01, 0.05, and 0.1. The hypothesis that the distribution of the observed data is Poisson with a rate parameter of 3.871549 cannot be rejected. If the bound of the rejection region is calculated for a α of 0.05, using the command qchisq(1-α, degrees of freedom), we get 16.91898, which is higher than the resulting statistic of 12.87262 and we reach the same conclusion, that the Poisson distribution hypothesis cannot be rejected.

```
> qchisq(.95,9)
[1] 16.91898
```

The chi-square goodness of fit test is one of the first tests proposed for this problem (Pearson, 1900) (Pearson, 1956).

Some of the advantages of this test are:

1. It is well-adapted when the probability distribution is not continuous and
2. The test statistic can be adapted when the distribution parameters are estimated from the sample, by the adjusted degrees of freedom.

Moreover, it needs large samples and it is recommended that the values of each one of the frequencies to be compared should be higher than 5.

Exponential Distribution

Exponential distribution is a probability distribution for continuous random variables, that is generally used for representations of time, T, and its interval of definition normally starts at zero and goes up to infinite.

Assuming the Bernoulli trial conditions that give rise to a Poisson distribution and that we want to model the time T taken for an interesting event to occur, while knowing that the average occurrence of events of interest per unit of time is λt, we want to find the probability of the time, t, that must elapse for an event to occur.

At first, using the Poisson distribution formula, the probability of an event of interest not occurring in a time t is calculated:

$$P\left(k=0\right)=\frac{\lambda t^{0}e^{-\lambda t}}{0!}=e^{-\lambda t} \tag{23}$$

Then the probability of the first event occurring before a time t is the complement:

$$P\left(T\le t\right)=1-e^{-\lambda t}=F\left(t\right)\text{for } t>0 \tag{24}$$

This is the cumulative distribution function for the variable time, the density function that corresponds to this CDF is:

$$f\left(t\right)=\lambda e^{-\lambda t}\text{ for } t>0 \tag{25}$$

The expected value, variance and coefficient of variation are, respectively:

$$\begin{aligned} \mathrm{E}\left[T\right]&=\frac{1}{\lambda}\\ V\left[T\right]&=\frac{1}{\lambda^{2}}\\ C\left[T\right]&=1 \end{aligned} \tag{26}$$

Value λ is usually called speed or rate while its inverse, 1/λ, the mean time between events of interest, is called a scale parameter, being the only parameter of the probability distribution.

It turns out then that for the same assumptions, the Poisson distribution, counting discrete events over time and the exponential distribution, measuring the time between events, the two distributions are related or else different models of the same phenomenon, depending on the form of observation.

- **Lack of Memory:** An exponential distribution similar to that of geometric distribution has the property of "lack of memory", which implies that the probability of an event of interest occurring in a time s, given that a time t has already past, is independent of the occurrence of t

$$P\left(T > t + s \middle| T > t\right) = \frac{P(T > t + s)}{P(T > t)} = \frac{e^{-\lambda(t+s)}}{e^{-\lambda t}} = q^{s} = P(T > s) \tag{27}$$

- **Software Use:** In the R statistic language, the probability functions for a random variable with an exponential probability distribution are calculated, the command dexp(t, λ) using for the probability density function f(t) and the command:

```
dexp(t, λ)
```

For the cumulative distribution function *F(t)=P(T≤t)* we use the command:

```
pexp(t, λ)
```

For example: We know that a certain event occurs on average 6 events per hour, λ=6, to find out the probability of an event occurring at 10 minutes = 1/6 hour, ¿P(T≤1/6)?, we use pexp(1/6,6).

```
> pexp(1/6,6)
[1] 0.6321206
```

What is the probability of an event taking more than one hour to happen? P(T>1)=1-P(T≤ 1)?

```
> 1-pexp(1,6)
[1] 0.002478752
```

The result can be compared with the probability for a Poisson random variable with λ=6 of no event *P(k=0)* occurring.

```
> dpois(0,6)
[1] 0.002478752
```

To get the inverse of the cumulative probability distribution function $F^{-1}(x)$, in other words, to obtain the value of quantile x, which corresponds to a cumulative probability F(x), we use the command qexp(F(x), λ), thus, for example, the quantile that corresponds to a probability of 0.6321 for an exponential distribution with parameter λ = 6, is:

```
> qexp(0.6321,6)
[1] 0.1666574
```

- **Random Data With Exponential Distribution:** To generate random data that follow an exponential distribution with parameter λ, we can use the formula:

$$t = -\frac{\log\left(rand\right)}{\lambda} \tag{28}$$

where: t is the random datum obtained with a parameter λ, and "log(rand)" is the natural logarithm of a uniform random number "rand" in the interval (0,1).

On a spreadsheet we can, for example, obtain for a value of λ=10: =-LN(RAND())/6, where LN() is the formula to calculate natural logarithm and RAND() is the formula for generating a pseudorandom number in the interval (0,1).

In the R statistical programming language, data are generated with exponential distribution using the command: rexp(sample size, λ). Thus, for example, to produce 10 data with exponential distribution and parameter λ=6, we have:

```
> rexp(10,6)
[1] 0.11226914 0.06602951 0.06195657 0.15581800 0.15237228
0.20956092
[7] 0.06313407 0.15031130 0.10889623 0.30828717
```

where [1] and [7] indicate the first and seventh elements of the vector of 10 elements with exponential distribution and parameter λ=6.

To estimate the scale parameter (1/λ), which is the inverse of the rate λ, in an exponential distribution, we use the arithmetic average of the observed data:

$$\bar{t} = \frac{\sum_{i=1}^{n} t_i}{n} \tag{29}$$

where: n is the sample size or number of data: $t_1, t_2, \ldots, t_n$, observed with an exponential distribution with rate λ, and their arithmetic average, $\bar{t}$ is calculated. This average will, in turn, be a random variable as it is as a function of the sum of n observed random variables.

This average is used as an estimator for the scale parameter (1/λ) for the exponential distribution. For example: For n=10 observations obtained from a process with exponential distribution, an average of 0.140437 was obtained Therefore that value is the estimator for the scale parameter. To estimate the rate parameter λ, we use the inverse 1/0.140437= 7.120631

Erlang Distribution

If we add "k" exponential random variables with the same parameter λ, the resulting sum will be a random variable with a distribution called Erlang distribution with parameters λ and k, the density function is:

$$f(t) = \lambda \frac{(\lambda t)^{k-1}}{(k-1)!} e^{-\lambda t} \text{ for } t > 0 \tag{30}$$

The cumulative probability distribution function is:

$$F(t) = 1 - \sum_{j=0}^{k-1} \frac{(\lambda t)^j}{j!} e^{-\lambda t} \text{ for } t \geq 0 \tag{31}$$

The expected value, variance and coefficient of variation are respectively

E[T]= k/λ; V[T] = k/λ^2; C[T]= $1/\sqrt{k}$.

Shape parameter k is an integer and λ (lower case lambda), rate or speed, is a real number, while Erlang distribution is a special case of the gamma distribution. In the case of Erlang distribution, shape parameter k is an integer.

Gamma probability distribution has a continuous number as its shape parameter and it is more common to find it defined with the scale parameter that corresponds to the inverse, 1/λ, of the rate parameter.

- **Software Use:** For the calculation of Erlang probabilities, we can use the gamma distribution commands that are to be found in the R programming language. Thus we use:

```
dgamma(x, k, λ).
```

whereas for the calculation of the cumulative probability distribution function F(x) we use the command: pgamma(x, k, λ).

To get quantile, x, relative to a cumulative probability F(x), the inverse cumulative function, we use:

```
qgamma(F(x), k, λ).
```

Thus, for example, for k =1, λ=5, (equivalent to an exponential with λ =5), we calculate for t = 1/5=0.2:

```
> dgamma(1/5,1,5)
[1] 1.839397
> pgamma(1/5,1,5)
[1] 0.6321206
```

For the inverse of the cumulative function in cumulative probability equal to 0.6321206, we get:

```
> qgamma(0.6321206,1,5)
[1] 0.2
```

- **Random Data for an Erlang Distribution:** To generate, on a spreadsheet, random data that have a Erlang distribution with parameters k and λ, we can use the formula:

$$t = -\frac{1}{\lambda}\log\left[\prod_{i=1}^{k} rand_i\right] \tag{32}$$

where: k and λ are the Erlang distribution parameters; "log" is the function for calculating the natural logarithm. The multiplication operator Π (capital pi) goes from 1 to k and is used to multiply "rand" data, uniform pseudorandom numbers in the interval [0,1].

The spreadsheet procedure for a random datum with Erlang distribution and parameters "k" and λ starts with the generation of "k" pseudorandom data in the interval [0,1]. Then we get the natural logarithm of its product. The result of the calculation of the logarithm is divided by –λ.

```
=LN(RAND()*RAND()*RAND)/(-6).
```

where LN() is the formula for calculating the natural logarithm, and RAND() is the command for generating pseudorandom numbers with uniform distribution in the interval (0,1).

In the R statistical programming language, random data with an Erlang distribution are generated using the command:

```
rgamma(sample size, k, λ).
```

For example. to generate a sample of 5 data with parameters k=12 y λ=6, we use:

```
> rgamma(5,12,6)
[1] 0.9944741 1.5353274 2.7795056 1.6129286 1.3697525
```

where [1] indicates that the first element of a vector of 5.

- **Relationship Between Erlang and Exponential Distributions:** Erlang distribution enables us to examine a property of a function derived from an exponential distribution. If we obtain a sample of n times, $t_1, t_2, \ldots, t_n$, observed with an exponential distribution with rate λ, and their arithmetic average, $\bar{t}$, is calculated, the resulting average will, in turn, be a random variable as it depends on the observed random variables. The arithmetic average is calculated

$$\bar{t} = \frac{\sum_{i=1}^{n} t_i}{n} \tag{33}$$

As has already been mentioned, the sum of n exponential random variables with parameter λ distributed according to an Erlang distribution that will result in "n" shape parameters and "λ" rate or speed parameters. Then, for the average t variable, the expected value shall be:

$$E\left[\bar{t}\right] = E\left[\frac{\sum_{i=1}^{n} t_i}{n}\right] = \frac{1}{n} E\left[\sum_{i=1}^{n} t_i\right] = \frac{1}{n}\left(\frac{n}{\lambda}\right) = \frac{1}{\lambda} = E\left[T\right] \tag{34}$$

The expected value of the average is equal to the expected value of the exponential distribution observed in the sample. The variance results in:

$$V\left[\bar{t}\right] = V\left[\frac{\sum_{i=1}^{n} t_i}{n}\right] = \frac{1}{n^2} V\left[\sum_{i=1}^{n} t_i\right] = \frac{1}{n^2}\left(\frac{n}{\lambda^2}\right) = \frac{1}{n}\left(\frac{1}{\lambda^2}\right) = \frac{1}{n} V\left[T\right] \tag{35}$$

The variance of the average is the variance of the observed exponential distribution divided by n, the number of observations. The standard deviation of the average is obtained using the square root of the variance and will be: $1/(\lambda n^{1/2})$.

In general for the average of a sample of n independently observed data $t_1, t_2, \ldots, t_n$, for a process or population with a probability distribution with mean E[T] and variance V[T], the arithmetic average obtained from the sample, $\bar{t}$, shall have a mean equal to that of the original population and a variance equal to the of the original population divided by n: $E\left[\bar{t}\right] = E\left[T\right]$;

$V\left[\bar{t}\right] = \frac{1}{n} V\left[T\right]$.

Confidence Intervals of Exponential Distribution

If X_1, X_2,…, X_n form a random sample of an exponential distribution with rate parameter = λ, then the variable $\lambda\bar{x}$ is distributed according to a gamma distribution in its Erlang variant with shape and rate parameters equal to the size of the sample: *n*.

If we wish to construct an interval, for λ, the rate parameter, with a confidence level with a value of 1- α, the two cut-off values are defined for a region that demarcates the probability 1- α within the distribution

Let E_{inf} be the value of the lower cut-off for the probability α/2 in the lower end of the Erlang distribution and E_{sup} be the cut-off point for 1-α/2 in the upper end. Then:

$$P\left(E_{\text{inf}} \leq \lambda\overline{x} \leq E_{\text{sup}}\right) = P\left(E_{\text{inf}} / \overline{x} \leq \lambda \leq E_{\text{sup}} / \overline{x}\right) = 1 - \alpha \tag{36}$$

The values of E_{inf} and E_{sup} can be obtained in the R statistical programming languages. As the Erlang Distribution is a case of gamma distribution, the commands used are the ones for E_{inf}:

```
=qgamma(0.025, forma=n, tasa=n).
```

and E_{sup}:

```
=qgamma(0.975, forma=n, tasa=n).
```

For example: For example, in order to calculate the confidence interval with value of 95%, α= 1-.95=.05, for a sample size n=10 for a population that is distributed according to an exponential distribution with rate parameter = 6.

```
> set.seed(1) #an initial seed is set for the randoms
> x<-rexp(10,6)# 10 data are generated with exponential distb
> mean(x)#the arithmetic average is calculated
[1] 0.140437
> qgamma(0.025,10,10)/mean(x) # is calculated the lower bound
[1] 3.414618
> qgamma(0.975,10,10)/mean(x) # the upper bound is calculated
[1] 12.16545
```

Thus the rate parameter λ is estimated in the interval [3.414618, 12.16545] with a 95% confidence level. It is then up to the person who is using the values to decide if the sample size gives them an interval with sufficient information.

It is important to have a strong certainty that the data model is effectively an exponential distribution and not another probability distribution model.

It is possible for an exact answer to be found for the wrong problem; so you should verify the data distribution model.

Central Limit Theorem and Normal Distribution

The Central Limit Theorem states that if S_n is the sum of n independent random variables, then the distribution of S_n is an approximation of the normal distribution given by the density function:

$$f\left(x\right)=\frac{1}{\sqrt{2\pi\sigma}}e^{-\left(x-\mu\right)^2/\left(2\sigma^2\right)} \tag{37}$$

where: f(x) is a normal probability density function; μ (lower case mu) is the mean or expected value of the function; σ (lower case sigma) is the standard deviation of function; π (lower case pi) is the constant 3.1415…, the value of the variable x goes from -∞ to +∞. Which is the entire real line. μ and σ, are respectively the location and scale parameters of the probability function.

The cumulative distribution function for the normal, F(x) is obtained by integration from -∞ up to value x. The cumulative function F(x) is calculated using numerical methods and the resulting values are reported in tables. Approximations are also calculated using formulas for their use in computer media. The values are reported for a single case of normal distribution. This is the standard normal distribution that have a mean equal to zero and a standard deviation equal to one.

- **Standard Variable:** A standard normal random variable is represented by the letter Z and is the result of transforming a random variable X with normal distribution, where X has an expected value or mean equal to μ and a standard deviation σ. The transformation consists of subtracting its mean from X and dividing the result by its standard deviation as shown below:

$$z=\frac{x-\mu}{\sigma} \tag{38}$$

The density function for the probability of a standardized normal random variable Z, is:

$$\left(z\right) = \frac{1}{\sqrt{2\pi}} e^{-\left(z\right)^2/2} \quad (39)$$

where $\phi(z)$, using ϕ (lower case phi), is how the normal density function for a standardized variable of the is represented; while the normal cumulative probability distribution for a standard variable z is represented by Φ(z), using a capital phi.

In general, for a random variable X with normal probability distribution, the expected value, variance and coefficient of variation are respectively: $E[X]=\mu$, $V[X]=\sigma^2$, $C[X]=\sigma/\mu$.

Another way of expressing the Central Limit Theorem is in respect of the arithmetic average, $\overline{x}$, "x-bar". If we take a sample of n observed data, $x_1, x_2, \ldots, x_n$, from a population with any probability distribution with mean μ, and standard deviation σ:

$$\overline{x} = \frac{\sum_{i=1}^{n} x_i}{n} \quad (40)$$

We say: if, $X_1, X_2, \ldots, X_n$ are independent random variables having the same distribution with mean μ and variance σ2. When n→∞, the probability distribution bound of the random variable $z = \frac{\overline{x} - \mu}{\sigma / \sqrt{n}}$ is the standard normal distribution.

Where z is a standardized variable, variable $\overline{x}$ is the arithmetic average of the sample of n observations that has a mean μ and a standard deviation $\overline{x}$

- **Software Use:** To get the probabilities for a normal variable in the R statistical programming language, we use the following commands: for the density function f(x) with mean μ and standard deviation σ, the command dnorm(x, μ,σ); to get the values of the cumulative probability distribution function F(x) we have the command:

```
pnorm(x, μ,σ).
```

For example, if, for a random variable with standard normal distribution, mean=0 and standard deviation =1, we want to calculate the value of the density function for the values -3, -2, -1, 0, 1, 2, 3, the calculation is as follows:

```
> x<-c(-3,-2,-1,0,1,2,3)
> dnorm(x,0,1)
[1] 0.004431848 0.053990967 0.241970725 0.398942280 0.241970725
0.053990967
[7] 0.004431848
```

It is worth mentioning that the function values for the number pairs at an equal absolute distance from the mean are similar. The calculated value of the function for -2 and 2 is 0.053990967.

For the cumulative distribution function, calculating the values for the set of numbers: -3, -2, -1, 0, 1, 2, 3; with mean = 0 and standard deviation =1, we get:

```
> x<-c(-3,-2,-1,0,1,2,3)
> pnorm(x,0,1)
[1] 0.001349898 0.022750132 0.158655254 0.500000000 0.841344746
0.977249868
[7] 0.998650102
```

In this case it must be pointed out that if 1-Φ(z) is calculated, we get Φ(-z), therefore 1-Φ(3)= Φ(-3)

```
> 1-0.998650102
[1] 0.001349898
```

To calculate the value of variable x with mean μ and standard deviation σ, corresponding to a cumulative probability, in other words, the quantile or the inverse cumulative function, we use the command in R:

```
qnorm(probability, μ,σ).
```

Thus, for example, for the standard variable, z, if ((z)= 0.975, the corresponding value of z is:

```
> qnorm(0.975,0,1)
[1] 1.959964
```

The values reported by the R language are calculated using formulas that numerically approximate the value of the functions. Nowadays the margin of error is fairly small and much better than the data available in printed tables.

- **Generating Normal Random Data:** One way of generating random data with standard normal distribution is by using the formula:

$$z_1 = \cos\left(2\pi * rand_1\right)\sqrt{-2\log\left(rand_2\right)}$$
$$z_2 = \sin\left(2\pi * rand_1\right)\sqrt{-2\log\left(rand_2\right)} \qquad (41)$$

where: Where z_1 and z_2 are independent random data with standard normal distribution, $rand_1$ and $rand_2$ are two pseudorandom numbers of uniform probability distribution with interval (0,1), "log" is the natural logarithm function. (Box & Muller, 1958)

To get normal random data, x, with mean and standard deviation σ. The data generated for a standard normal z are used: z is multiplied by the standard deviation and the mean is added to the product:

$$x = \mu + \sigma z \qquad (42)$$

Thus, for example: A datum z is obtained on a spreadsheet by using:

```
=COS(2*PI()*RAND())*SQRT(-2*LN(RAND()))
```

If we want to generate data with, for example, a mean of 50 and a standard deviation of 5, the commands are:

```
=50+5*COS(2*PI()*RAND())*SQRT(-2*LN(RAND()))
```

To generate random data with normal distribution with mean μ and standard deviation σ in the R statistical programming language, we use the command:

```
rnorm(sample size, μ, σ).
```

For example: to generate a vector with 7 normal random data with mean =50 and standard deviation =5, we have:

```
> rnorm(7,50,5)
[1] 47.71827 50.67851 52.51401 50.13377 56.05774 48.62948
47.80221
```

- **Estimators for Normal Parameters:** To estimate the normal distribution parameters, the arithmetic average of "n" observations of a random sample taken from the population of interest, denoted as the "sample mean" or "x-bar", $\overline{x}$ are used for the mean; "variance sample 1", denoted as S^2, is used for the variance; while the estimators are:

$$\overline{x} = \frac{\sum_{i=1}^{n} x_i}{n} \qquad (43)$$

$$S^2 = \frac{\sum_{i=1}^{n}\left(x_i - \overline{x}\right)^2}{n-1} = \frac{\sum_{i=1}^{n}\left(x_i\right)^2 - n\overline{x}^2}{n-1}$$

where: x_i (i=1,n) are the n independently observed and random data of a population; $\overline{x}$ is the arithmetic average of the n data while also being the estimator of the mean μ; while S^2 is the sample variance of the n observed data. The second part of this formula is an equivalent alternative, and easier of calculate numerically, for the first part of the formula. The standard deviation of the sample is estimated obtaining the square root of the variance sample 1 and is denoted as: s.

Confidence Intervals for the Mean of a Normal Distribution

To get a confidence interval for a normal distribution when the two distribution parameters are unknown and we have to use the arithmetic average $\overline{x}$ for mean μ and sampling variance s^2 to estimate the variance of the population σ2, we use the random variable defined as:

$$\frac{\overline{x} - \mu}{s / \sqrt{n}} \qquad (44)$$

This has student's t distribution with parameter n-1 degrees of freedom to construct a confidence interval with a probability 1-α of it taking a value between the cut-off values: ($t_{\alpha/2,n-1}$, $t_{1-\alpha/2,n-1}$). Because of the symmetry of the student's t distribution, the two bounds are equal in absolute value: $|t_{\alpha/2,n-1}| = |t_{1-\alpha/2,n-1}|$= cut-off value, thus we have:

$$P\left(t_{\alpha/2;n-1} < \frac{\overline{x}-\mu}{s\,/\,\sqrt{n}} < t_{1-\alpha/2;n-1}\right) = 1-\alpha \tag{45}$$

That can be converted into an interval equivalent to estimating the mean μ:

$$P\left(x - |\,t_{\alpha/2;n-1}\,|\frac{s}{\sqrt{n}} < \mu\left\langle\overline{x} + |\,t_{\alpha/2;n-1}\right|\frac{s}{\sqrt{n}}\right) = 1-\alpha \tag{46}$$

To get the limit value for student's t, $t_{\alpha/2;n-1}$ in the R statistical programming language, we can use the command:

qt(alfa/2, n-1).

- **For Example:** Using the R statistical programming language, 70 data are generated with mean =50 and standard deviation =5, and a 95% confidence interval is constructed, alpha=5%, for the mean:

```
> set.seed(123) # a seed is set for the pseudorandoms
> x<-rnorm(70, 50,5)# n=70 normal data are generated, μ=50, σ=5
> xt<-mean(x); S<-sd(x);T<-abs(qt(0.025,69))# calculations
> xt;S;T # average, standard deviation, |t_(alpha/2),n-1|
[1] 50.36913 # arithmetic average
[1] 4.537208 # standard deviation
[1] 1.994945 # |t_(alpha/2),n-1|
> Lower<-xt-T*S/sqrt(70) # calculation of lower bound
> Upper<-xt+T*S/sqrt(70) # calculation of upper bound
> Lower; Upper
[1] 49.28727 # lower bound
[1] 51.45099 # upper bound
```

Thus the result was that, for a sample of 70 observed data of a normal distribution, the mean was estimated as 50.36913; the standard deviation was estimated as 4.537208 and the cut-off value calculated with a student's t of 69 degrees of freedom to give us an extreme with 0.025 of probability in 1.994945. The bounds for a confidence level of. 95 =1- 0.05 resulted in an estimation of the mean as being between 49.28727 and 51.45099.

When the standard deviation is known or when the size of the sample very large, the estimation of the confidence interval uses normal distribution instead of student's t distribution, and instead of using the command qt(alpha/2, n-1), we can use the command qnorm(alpha/2) which gives us the cut-off value in a normal standard variable, z, for a cumulative probability of alpha/2.

Confidence Intervals for Normal-Distribution Variance

To construct the 1- α confidence interval for the variance, σ^2, for data with a normal distribution based on a sample of n randomly and independently observed data and using sample variance s^2, the calculation is done from the variable defined as:

$$\frac{\left(n-1\right)s^2}{\sigma^2} \tag{47}$$

This variable is distributed like a chi-squared variable, χ^2, with n-1 degrees of freedom. In the formula, "n" is the size of the sample; the number of observed data; s^2 is the sample variance calculated from the n data and σ^2 is the variance of the population we are trying to estimate.

The interval for a confidence level of 1-α is constructed on the basis of two cut-off bounds that indicate a region with 1-α probability and $\alpha/2$ extremes; in other words:

$$P\left({}^{2}_{\alpha/2,n-1} < \frac{\left(n-1\right)s^2}{\sigma^2} < {}^{2}_{1-\alpha/2,n-1}\right) = 1-\alpha \tag{48}$$

That is equivalent to:

$$P\left(\frac{\left(n-1\right)s^2}{{}^{2}_{1-\alpha/2,n-1}} < \sigma^2 < \frac{\left(n-1\right)s^2}{{}^{2}_{\alpha/2,n-1}}\right) = 1-\alpha \tag{49}$$

The chi-square variable, ${}^{2}_{\alpha/2,n-1}$, can be obtained by using the command:

qchisq(α/2, n-1).

In the R statistical programming language.

- **For Example:** Using the R statistical programming language, 70 data are generated with mean =50 and standard deviation =5. Then a 95% confidence interval is constructed for the variance:

```
> set.seed(123) # a seed is set for the pseudorandoms
> x<-rnorm(70, 50,5)# n=70 normal data are generated, μ=50, σ=5
> var(x)
[1] 20.58626
> 69*var(x)/qchisq(.025,69)
[1] 29.63958
> 69*var(x)/qchisq(.975,69)
[1] 15.1343
```

For 70 observed data, the variance is estimated as 20.58626 with a 95% confidence interval of (15.1343, 29.63958). If we consider the standard deviation, this is used to get the square root of the values obtained for the variance, which gives us:

```
>sqrt(c(var(x),69*var(x)/qchisq(.975,69),69*var(x)/
qchisq(.025,69)+))
[1] 4.537208 3.890283 5.444224
```

An estimated standard deviation of 4.537208 is obtained with a 95% confidence interval of (3.890283, 5.444224).

Goodness of Fit Tests

Calculations based on a probability distribution model with its established parameters depend on the election of these models, so we need to check that the model is being used correctly and reasonably. To appraise the probability distribution model, we use goodness of fit tests: Goodness of Fit Tests test the hypothesis: these data follows the given probability distribution. Let us assume that we have a sample of *n* observed random data for a process or population: $x_1, x_2, \ldots, x_n$ and we want to prove or find out what probability distribution model is the most suitable for describing them.

Quantile Method

In this method, the observed quantiles compared with the ones that the data should, in theory, follow, assuming that the observed data follow the proposed probability distribution.

If there is a cumulative probability distribution function, we say that $F(x)=P(X\leq x)$ and function assigns a cumulative probability value to the value of x. A quantile makes the inverse function assign the value of x to a cumulative probability. The quantile (p) is the value $x=F^{-1}(p)$, until a cumulate quantity *p* of probability is found.

Expected or theoretical quantiles should also be generated for the distribution that we want to compare with the values obtained from a sample.

- **Procedure:** Starting from the existence of a sample of n data that have been ordered $x_1 \leq x_2 \leq \ldots \leq x_n$.

```
> set.seed(3) # a seed for the random numbers is started
> x<-rexp(50,10) # n=50 data are generated with λ=10
> x<-sort(x) # the data are ordered
```

Then a sequence is formed of integer numbers from 1 to *n*. This sequence is the indices of the sample, where *n* corresponds to the number of data in the sample and is the number of values to be calculated.

```
> index<-1:50
```

Each of the integer values of the indices, known as *i*, is assigned a cumulative probability = (i-03)/(n+0.4).

```
> pr < - (index-0.3/(50.4)
```

For every probability assigned for the indices, theoretical quantiles, q_1, q_2, …, q_n, are calculated in accordance with the distribution from which the data are thought to have come. The calculated quantiles, q_i, are such that $P(X\leq q_i)=(i-0.3)/(n+0.4)$. We need to test an exponential distribution with $\lambda=10$.

```
> qteor<- qexp(pr,10)
```

The correlation between the two sets of data is calculated. A high correlation implies that the model could be a good option. A hypothesis is tested to check whether the correlation is zero. If the significant p value is less that alpha, the null hypothesis that the correlation is zero is rejected.

```
> cor.test(x,qteor)
Pearson's product-moment correlation
data: x and qteor
t = 86.4365, df = 48, p-value < 2.2e-16
alternative hypothesis: true correlation is not equal to 0
95 percent confidence interval:
0.9943439 0.9981940
sample estimates:
cor
0.9968031
```

For an alpha of 0.05, the P value is smaller. The hypothesis that the correlation is zero is rejected..

- **Drawing a Plot:** To get a plot with the data from the above example:

```
> plot(x~qteor) # draw the sample data versus theoretical data
> abline(lm(x~qteor),col=2) # add a straight red line.
```

Kolmogorov-Smirnov Test

The Kolmogorov-Smirnov test is a nonparametric test that lets us compare two cumulative probability distributions for continuous variables. A statistic called D is calculated that estimates the longer vertical distance between the two cumulative distribution functions and if the statistic exceeds a critical value obtained for the distribution of distances, the null hypothesis that both distribution functions are equal is rejected. (Kolmogorov, 1933) (Smirnov, 1948)

The test is used in two ways:

1. For comparing two empirical cumulative distribution functions
2. For comparing an empirical distribution function with a theoretical cumulative distribution function.

In any case, the test assumes independence between the two distribution functions. Thus, if the theoretical cumulative distribution function is proposed

Figure 1. Plot that shows quantiles from observed sample vs. theoretical quantiles, assuming a distribution

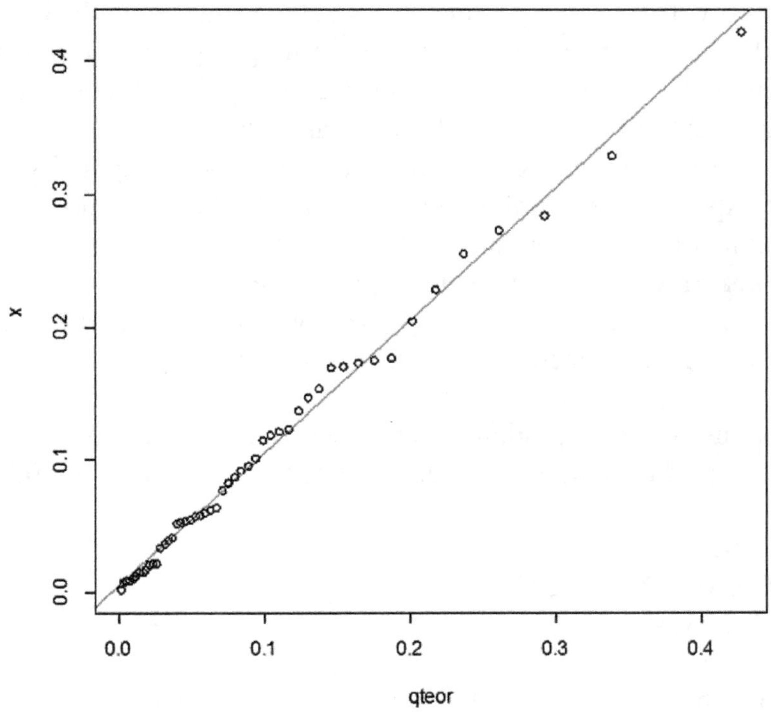

using estimated parameters for the data of the empirical distribution function, the condition of independence is broken.

When the Kolmogorov-Smirnov test is used to compare an empirical cumulative distribution function with a theoretical cumulative distribution function, we should start with parameters that are independently postulated for the theoretical part.

One direct application of the Kolmogorov-Smirnov test is presented to explain a result that is known in the literature as Burke's Theorem.

- **Burke's Theorem:** Burke's Theorem says that, for a queuing system, with a single queue, without departures, for any number of servers and independent times between arrivals with exponential distribution, the time intervals between successive outputs are independently and exponentially distributed with a parameter equal to the parameter for the exponential distribution of times between arrivals. (Burke, 1956)

- **Simulation of Queuing System:** In order to visualize this, we carry out a simulation by generating exponentially distributed times between arrival with a known parameter, λ =6, the we generate the behavior of 100 arrivals arriving in the system, forming a queue, if necessary, and being served with exponential service times with parameter μ= 12. A log is kept of the 100 times between arrivals and the 100 output times for the corresponding system. The arrival and output distributions are expected to be similar when compared. The test is done using the Kolmogorov-Smirnov test for the comparison of two distributions.
- **Generation of Times:** Using a spreadsheet with an initial seed to get reproducible results, and using pseudorandom numbers produced by the congruential method, the data is shown in Table 6.

In column B, pseudorandom numbers with uniform distribution (0,1) are being generated using the congruential method, using a root of 30,001:

A2: =30001
A3: =MOD(48271*A2;2^31-1)
B3: =A3/(2^31-1)

In B3 a uniform pseudorandom number is produced in the interval [0,1].

Columns C and D repeat the same process as A and B but this time starting from a root of 20,001. Data with a uniform distribution (0,1) are also obtained in column D at the end.

In column E, exponential time data with parameter λ=6, (E1) are generated, using the uniform data from column B:

E1: = 6
E3: =-LN(B3)/$E$1

Table 6. Segment of spreadsheet showing the start of the simulation of arrival times, service times and output times for a queuing system

	A	B	C	D	E	F	G	H	I
1					6	12			
2	30001		20001		ta	ts	clock a	clock d	td
3	1.45E+09	0.6744	9.65E+08	0.4495	0.0657	0.0666	0.0657	0.1323	0.1323
4	1.26E+08	0.0585	1.58E+09	0.7340	0.4731	0.0258	0.5388	0.5645	0.4323

E3 is copied for the rest of the column. The 100 data in E simulate the times between arrivals with a rate parameter λ=6 arrivals/unit of time.

In column F, data for exponential service times are generated with parameter λ=12 services /unit of time, (F1), using the uniform data from column D:

F1: =12
F3: =-LN(D3)/$F$1

F3 is copied for the rest of column F. These 100 data simulate the times between arrivals with a rate parameter λ=12.

Column G -Clock A logs the arrival time for each unit that enters.

G3=E3
G4=G3+E4

In cell G3, it starts with the time of the first arrival, =E3.

Starting from cell G4, the last time is accrued to the corresponding time between arrivals, =G3+E4, and the rest of the column copies the formula.

Column H –Clock D logs the output time of each unit served:

H3: = G3+F3
H4:= F4+MAX(H3,G4).

The output time for the initial unit in H3 is the sum of the arrival time, G3, plus service time, F3.

The output time of the next unit is calculated as the sum of the service time, F4, plus the maximum between the output times of the previous unit, H3; and the arrival time of the present unit, G4. The rest of the column copies the formula.

Finally, column I –td calculates the output time for each unit:

I3: = H3,
I4: = H4-H3

In I3 this is obtained with the log of the output clock of the first unit served, H3.

In I4 the time between outputs is obtained by the difference between the clock value for the output of the present unit, H4, and the output time of the last unit, H3. The rest of the column copies this formula.

The columns are copied until they generate 100 data. The column of the times between arrivals, (ta) column E, and the column with the times between outputs, (td) **column I**, that according to the Burke's theorem is expected to have similar exponential distributions are worth looking at. The Kolmogorov-Smirnov test will be used for two samples to check the similarity.

- **Descriptive Statistics and Graphs:** The data can be copied into the R program from the spreadsheet. Open the R program and give it the instruction to read from the "clipboard", assigning a variable called Ta, without running it yet:

```
> Ta<- read.table(file = "clipboard", sep="\t", header=TRUE)
```

On the spreadsheet, select and copy the column with the arrival times, ta (column E in this exercise), including the heading. This goes in the interval of cells E2 to E102. Go back to the R program and run the assignment command.

The command to read from the "clipboard" is repeated in the R program, but now assigning a variable called Td. Do not run yet:

```
> Td<- read.table(file = "clipboard", sep="\t", header=TRUE)
```

On the spreadsheet, select and copy the column with the arrival times, td (column I in this exercise), including the heading. This goes in the interval of cells I2 to I102. Go back to the R program and run the assignment command.

The two read lists, Ta and Td, are attached to the search path of the R program, and a basic statistics summary is calculated for the vectors ta (times between arrivals) and td (times between outputs):

```
> attach(Ta)
> summary(ta)
Min. 1st Qu. Median Mean 3rd Qu. Max.
0.001508 0.035150 0.124100 0.179200 0.289300 0.979700
> attach(Td)
> summary(td)
Min. 1st Qu. Median Mean 3rd Qu. Max.
0.0006237 0.0448700 0.1178000 0.1794000 0.2717000 0.9868000
```

Note that the calculated statistics are very approximate in value. The inverse of the calculated averages are for ta and td respectively: 5.58 and 5.573.

- **Histogram:** A histogram is drawn to observe the behavior of the distribution of the 100 data generated and stored in vector ta:

```
> hist(ta)
```

- **Quantile Plot:** A quantile plot is drawn to compare the arrival data on vector ta with the theoretical data, assuming an exponential distribution with rate λ=6:

```
> t.teo<-rexp(100,rate=6) # 100 exponential data are generated
with rate =6
> qqplot(t.teo,ta,main="QQ-plot distr. exponential") # the
quantile plot is drawn to compare the theoretical times, t.teo,
with ta.
> abline(0,1, col=2) # a line is added at 45° degrees
```

- **ECDF Graph:** Now we draw a graph to compare the cumulative distribution functions for the empirical arrival data in ta and the theoretical values for an exponential distribution with a rate equal to 6

Figure 2. Histogram of the 100 arrival times generated. This shows a skewed distribution, as an exponential distribution appears.

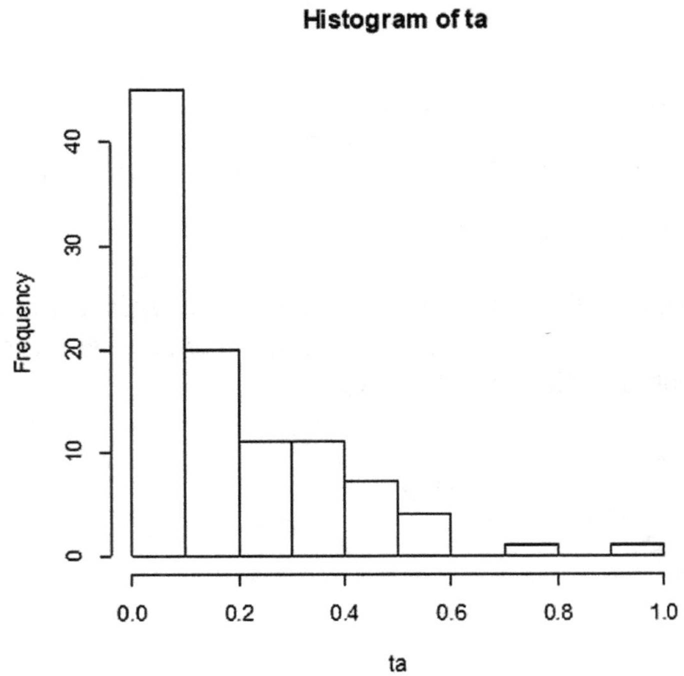

Figure 3. Quantile plot comparison using the qqplot command, with a straight line being added at an angle of 45°

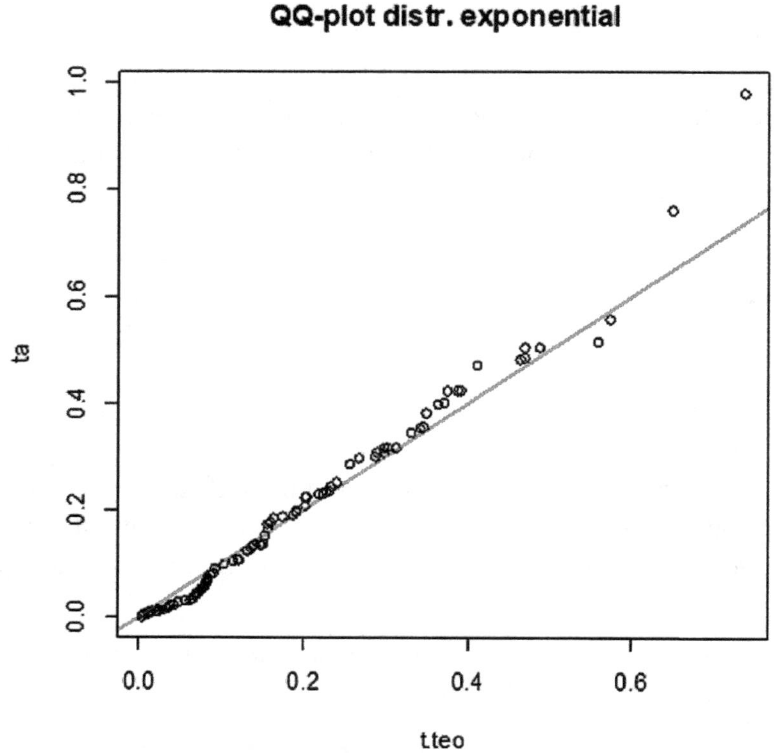


```
t<-seq(0,4*0.1792,0.1) # a sequence is generated of times
between 0 and four times the average of ta.
> plot(t,pexp(t,rate=6),type="l", main="ECDF and Exponential
CDF")
> plot(ecdf(ta),add=TRUE)
```

- **Goodness of Fit Test:** A Kolmogorov-Smirnov test is run to compare the distribution of the 100 arrival times, on vector ta, with an exponential distribution with parameter λ=6:

```
> ks.test(ta, "pexp", 6)
One-sample Kolmogorov-Smirnov test
data: ta
D = 0.083235, p-value = 0.4925
alternative hypothesis: two-sided
```

Figure 4. Cumulative distribution graphs. The theoretical cumulative distribution are represented by a line and the observed empirical cumulative distribution are represented by points. The Kolmogorov-Smirnov test estimates the longest vertical distance between the empirical cumulative distribution (stepped figure) and the theoretical distribution cumulative (unbroken line).

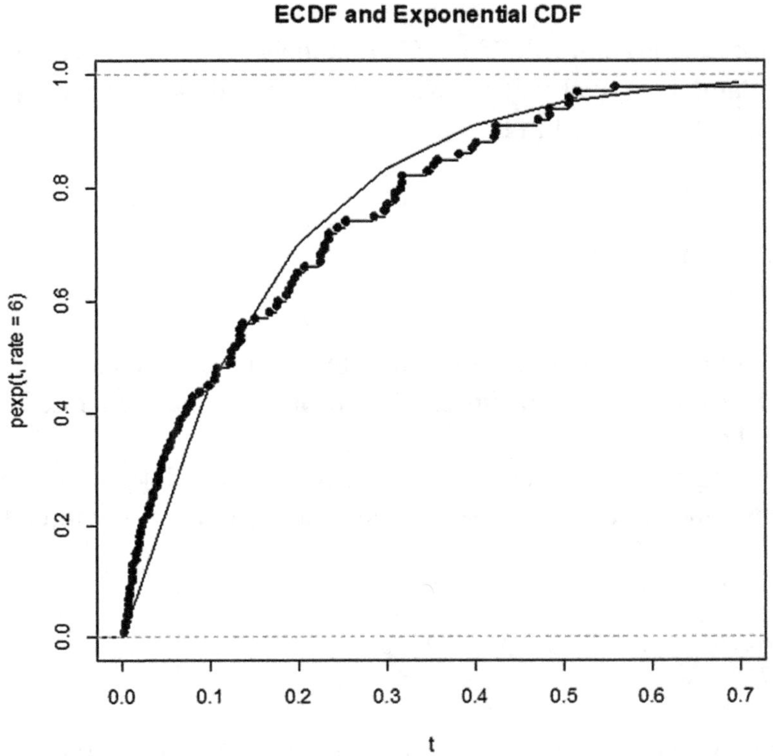

We cannot reject the null hypothesis that the data in ta follow an exponential distribution with rate λ=6; the P value of 0.4925 is much higher than the usual alpha values of 0.01, 0.05 and 0.10.

- **Comparison of Distributions:** Now the Kolmogorov-Smirnov test is run to compare the distributions of the two samples:

```
> ks.test(ta,td)
Two-sample Kolmogorov-Smirnov test
data: ta and td
D = 0.12, p-value = 0.4676
alternative hypothesis: two-sided
```

There is not enough evidence to reject the null hypothesis of the Kolmogorov-Smirnov test. Ho: the two continuous distributions are equal. So we are left with the hypothesis that both continuous distributions are similar. The reported P value is 0.4676, which is a very high value in respect of the significant values generally being used, such as 0.01, 0.05 and 0.1.

- **Confidence Interval:** If a 95% confidence interval is calculated for the lambda arrival value, assuming that the data have an exponential distribution, we get in R:

```
> qgamma(0.025,100,100)/0.179200
[1] 4.540401
> qgamma(0.975,100,100)/0.179200
[1] 6.725946
```

Assuming that the probability distribution is exponential, the lambda value for arrival time is estimated based on 100 data generated between 4.540 and 6.726.

Assuming that the output times have an exponential distribution, their 95% confidence interval is: The 95% confidence interval is estimated in:

```
> qgamma(0.025,100,100)/0.1794000
[1] 4.53534
> qgamma(0.975,100,100)/0.1794000
[1] 6.718447
```

In other words, the 95% confidence interval for the output rate is between 4.534 and 6.718.

The Kolmogorov-Smirnov test is a nonparametric test that enables us to compare two independent samples of continuous variables and test to see if they have the same distribution. The Kolmogorov-Smirnov test needs large sample sizes. It can be used to compare a sample of continuous variables with the theoretical values of a supposed distribution.

However, in general, because of the necessity of large samples, the Anderson-Darling test is thought to be better than the Kolmogorov-Smirnov test. The Anderson-Darling test is also nonparametric but can use smaller sample sizes. The Anderson-Darling test can, moreover, use the parameters estimated from the sample for the theoretical distribution assumed in the test. In the Kolmogorov-Smirnov test the use of estimators based on the sample invalidates the assumption of independence between the two sets to be compared.

Anderson-Darling Test

The Anderson-Darling test is a nonparametric test that performs a statistical hypothesis test with the null hypothesis Ho: the data of a sample follow a distribution. The test is applied to univariate continuous random variables and variants of the test have been developed for different distributions. (Anderson & Darling, 1952) (Anderson & Darling, 1954).

The A^2 statistic that calculates the Anderson-Darling test is:

$$A^2 = -n - \sum_{i=1}^{n} \frac{(2i-1)}{n} \left[\log F\left(x_i\right) + \log\left(1 - F\left(x_{n+1-i}\right)\right) \right] \quad (50)$$

where: n is the size of the observed data sample, x_i (i=1,n) is the observed datum in i-th place of the ordered set of observed data $x_1 < x_2 < \ldots < x_n$; "F" is the proposed continuous Cumulative Distribution Function and "log", the function natural logarithm.

The null hypothesis is that the data follow the proposed cumulative distribution F. The hypothesis is rejected if the test statistic is higher than a critical value. There are calculation algorithms for different specific distributions. This implies the use of computer calculations.

- **Installing a Package:** In order to perform the Anderson-Darling test in the R statistical programming language, an additional package needs to be installed. We will use the goftest package (Faraway, Marsaglia, Marsaglia & Baddeley, 2015). The instructions for its installation, once we have access to the web, are to give the command:

```
> install.packages("goftest")
```

A request will appear asking us to specify the site we want to download the package from. Once we have indicated the site and the command has been run, the following command is given:

```
> library(goftest)
```

The syntax of the function for running the Anderson-Darling test has a similar structure to that of the Kolmogorov-Smirnov test:

```
ad.test(x, "pdist", parameters).
```

Where: “x” is the data vector we wish to test; “pdist” is the name in R of the proposed cumulative distribution declared between inverted commas and “parameters” is the list separated by commas of the parameters needed for the distribution.

- **Goodness of Fit Test:** Thus, for example, in the case of a queue simulation that was employed to present the Burke’s Theorem, the Anderson-Darling test is run for the 100 arrival time data on vector ta, to check whether they have an exponential distribution with rate λ=6, compared with a 5% alpha:

```
> ad.test(ta, "pexp", rate=6)
Anderson-Darling test of goodness-of-fit
Null hypothesis: exponential distribution
with parameter rate = 6
data: ta
An = 1.4536, p-value = 0.188
```

The null hypothesis cannot be rejected. Ho: the data follow an exponential distribution with rate =6”. The P value is higher than alpha: 0.188>0.05

The command could also be written thus:

```
> ad.test(ta, "pexp",6)
```

The numeric result is exactly equal, the explanation is limited to:

```
Anderson-Darling test of goodness-of-fit
Null hypothesis: exponential distribution
data: ta
An = 1.4536, p-value = 0.188
```

- **Test With Gamma:** Also, remembering that exponential distribution can be seen as a case of gamma distribution, when the shape parameter of gamma is equal to one, it can be done as follows:

```
> ad.test(ta, "pgamma",shape=1,rate=6)
Anderson-Darling test of goodness-of-fit
Null hypothesis: Gamma distribution
with parameters shape = 1, rate = 6
data: ta
An = 1.4536, p-value = 0.188
```

With similar results.

- **Test With Normal:** If we do the test for a normal distribution with a mean equal to 1/6 and standard deviation equal to 1/6, for an alpha of 5%, we have:

```
> ad.test(ta, "pnorm",mean=1/6,sd=1/6)
Anderson-Darling test of goodness-of-fit
Null hypothesis: Normal distribution
with parameters mean = 0.166666666666667, sd =
0.166666666666667
data: ta
An = 3.8614, p-value = 0.01024
```

The "P" value is less than 0.05. The hypothesis that the data have a normal distribution with mean = 1/6 and standard deviation =1/6 is rejected.

The Anderson-Darling test does not let us compare two sets of data to see if they have the same distribution. We would have to test each set of data separately against a theoretical distribution that they are assumed to have in common. So we do the following test for the output times, on vector td:

```
> ad.test(td,"pexp",rate=6)
Anderson-Darling test of goodness-of-fit
Null hypothesis: exponential distribution
with parameter rate = 6
data: td
An = 0.62273, p-value = 0.6264
```

It turns out that the data of output times give a P value in the Anderson-Darling test, higher than the commonly-used alpha values of 0.01, 0.05 and 0.1. So, in conclusion, the hypothesis that the output times have an exponential distribution with a rate equal to 6 cannot be rejected

- **Use of Methods for Testing Goodness of Fit:** We have presented three goodness-of-fit hypothesis tests. The comparison of quantiles with quantiles is more by way of an illustration to help us understand the goodness of fit between two distribution functions.

The chi-squared test compares observed frequencies with expected frequencies and can be used in the case of distributions that are not continuous If the parameters of the theoretical function that generates the expected frequencies are estimated, the degrees of freedom decrease in accordance with the number of calculated parameters. However the chi-squared test needs large samples and any frequency that is higher than 5 is recommended. As the observed frequencies often come from histograms and the frequencies depend on the size and number of the classes chosen, this entails a certain loss of information.

The Kolmogorov-Smirnov test enables us to compare two cumulative distribution functions based on the longer vertical distance between the two distribution functions. This test is used in the case of univariate and independent continuous distributions. While the use of empirical cumulative distribution functions is an advantage over the use of frequencies that come from histograms, its application is limited by its assumptions. It also has some difficulties when value of the data of the two functions match.

The Anderson-Darling test compares ordered observed data with a theoretical cumulative distribution function and gets a statistic, taking advantage of all the information available in the data. It is used for comparing univariate continuous variables in general, in the case of univariate continuous functions. The Anderson-Darling test is recommended for testing the fit of empirical data with a theoretical distribution function is preferred over the other two options (Stephens, 1974).

REFERENCES

Anderson, T. W., & Darling, D. A. (1952). Asymptotic theory of certain “goodness-of-fit” criteria based on stochastic processes. *Annals of Mathematical Statistics*, *23*(2), 193–212. doi:10.1214/aoms/1177729437

Anderson, T. W., & Darling, D. A. (1954). A Test of Goodness-of-Fit. *Journal of the American Statistical Association*, *49*(268), 765–769. doi:10.1080/01621459.1954.10501232

Box, G. E. P., & Muller, M. E. (1958). A Note on the Generation of Random Normal Deviates. *Annals of Mathematical Statistics*, *29*(2), 610–611. doi:10.1214/aoms/1177706645

Burke, P. J. (1956). The Output of a Queuing System. *Operations Research*, *4*(6), 699–704. doi:10.1287/opre.4.6.699

Faraway, J., Marsaglia, G., Marsaglia, J., & Baddeley, A. (2015). *Goftest: Classical Goodness-of-Fit Tests for Univariate Distributions*. R package version 1.0-3. Retrieved from https://CRAN.R-project.org/package=goftest

Grinstead, C. M., & Snell, J. L. (2012). *Introduction to probability*. American Mathematical Soc.

Kolmogorov, A. (1933). Sulla determinazione empirica di una legge di distributione, *Giornale dell'. Istituto Italiano degli Attuari*, *4*, 83–91.

Pearson, K. (1900). On the criterion that a given system of deviations from the probable in the case of a correlated system of variables is such that it can be reasonably supposed to have arisen from random sampling. *Phil. Mag.*, *5*(50), 157-175.

Pearson, K. (1956). *Karl Pearson's Early Statistical Papers*. Cambridge University Press.

R Core Team. (2015). *R: A language and environment for statistical computing*. R Foundation for Statistical Computing. Retrieved from https://www.R-project.org/

Rutherford, E., Geiger, H., & Bateman, H. (1910). LXXVI. The probability variations in the distribution of α particles. *Philosophical Magazine Series 6*, *20*(118), 698–707. doi:10.1080/14786441008636955

Smirnov, N. (1948). Table for estimating the goodness of fit of empirical distributions. *Annals of Mathematical Statistics*, *19*(2), 279–281. doi:10.1214/aoms/1177730256

Stephens, M. A. (1974). EDF statistics for goodness of fit and some comparisons. *J. Amer. Stat. Assn*, *69*(347), 730–737. doi:10.1080/01621459.1974.10480196

Chapter 3
Single Stage Systems Markovian Case

ABSTRACT

In Chapter 3, the authors show the expressions of queueing theory for Markovian systems with a single stage. The chapter begins with definitions of stochastic processes and Markov chains; then; they present the models for calculating the work in process and the cycle time of systems with a single server, multiple servers, and systems with restriction on queue size. Later, the chapter explores heuristic rules to estimate the capacity of a system. The chapter ends with the monetary analysis of the system and the optimum selection of capacity.

STOCHASTIC PROCESSES

This is a set X of random variables that appear as a parameter t, whose elements belong to a set T, passes. A stochastic process is defined as X (t). In applications of stochastic processes for modeling real systems it is very common for set T to refer to a time sequence and each element t of the set to represent (discrete or continuous) instants of time within the same sequence; therefore the elements of the time sequence are $T= \{0, 1, 2,\ldots\}$. Furthermore set X contains the possible events or results: taking a card from a deck, throwing the dice, the result of an experiment are examples of experimental events. Other situations are, for example, the number of customers that arrive

DOI: 10.4018/978-1-5225-5264-2.ch003

at a box-office or leave a row at any given time or the failure of a piece of equipment in a particular moment.

A stochastic process is defined by:

1. The possible events of set *X*, which are known as a set of possible states or just as states.
2. The elements of set *T*.
3. The relationship or connection that exists between every possible state.

The relationship between the variables of a stochastic process is as shown in Table 1.

Markov Chains

A Markov chain (or process) $X(t)$ is a stochastic process with the property that if an event X_t arises, then all other possible future events X_t for the following instants of time $s > t$ are not affected by any past actions or events $u < t$ that brought the system to state X_t. (Taylor & Karlin, 1984)

In terms of probability, a stochastic process is one where, when a system is found in a particular state, the probability of getting the next action or result is not affected by the events or the path followed in the instants before the current state.

A Markov chain (Figure 1) is a process where set *X* has a number of countable and finite elements and time happens in (discrete or continuous) intervals of set $T= \{0, 1, 2,\dots\}$; in formal terms of conditional probability:

$$\begin{aligned} P\left(X \leq x\right) &= F_X\left(x\right) = \int_{-\infty}^{x} f\left(x\right) dx \text{ X continuous} \\ P\left(N \leq k\right) &= F_N\left(k\right) = \sum_{k=-\infty}^{k} f\left(k\right) \text{N discrete} \end{aligned} \tag{1}$$

Table 1. Classification of stochastic processes accordingly to their variables

Time	States	
	Discrete	Continuous
Discrete	Stochastic chain at discrete time	Stochastic process at discrete time
Continuous	Stochastic chain at continuous time	Stochastic process at continuous time

Figure 1. Markov Chain with 4 states and transition probabilities p and 1-p

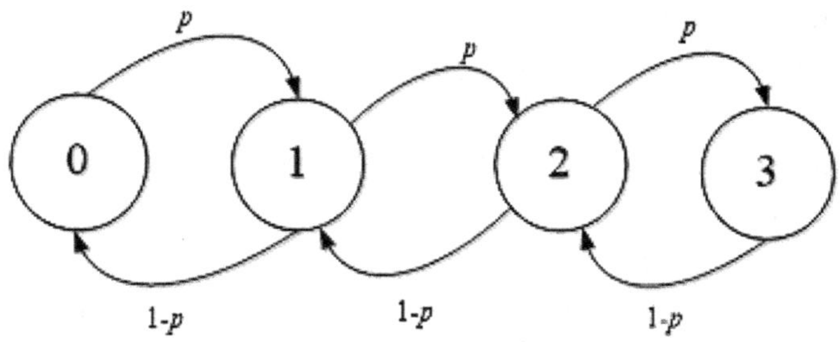

For all the instants of time n and all the states $i_0,\ldots, i_{n-1}, i, j$. Probability X_{n+1} of going to state j, given that in instant X_n the system is in state i, is known as transition probability $p_{ij}^{n,n+1}$:

$$E[X] = \int_{-\infty}^{\infty} xf(x)\,dx \text{ X continuous}$$
$$E[N] = \sum_{k=-\infty}^{\infty} kf(k) \text{N discrete} \quad (2)$$

Given that probabilities $p_{ij}^{n,n+1}$are independent of instant n, the transition probability is simply written as p_{ij} and, for a better representation, they are fitted into a square matrix. Each entry of the matrix indicates the probability of going from one state to another:

$$V[X] = \int_{-\infty}^{\infty} (x - E[X])^2 f(x)\,dx \text{ X continuous}$$
$$V[N] = \sum_{k=-\infty}^{\infty} (k - E[N])^2 f(k) \text{N discrete} \quad (3)$$

The matrix P is called a Transition Matrix or Markov Matrix. Every line of matrix P is the probability distribution for going from a state i to a state j, given that the system is in state i. If the set of states is finite, then the matrix is a matrix square with a finite number of rows and columns equal to the number of states of the system. Each line has the following properties:

$$C\left(X\right)=\frac{\sqrt{V\left[X\right]}}{E\left[X\right]}\text{ For } i,j=0,1,2,\ldots \quad (4)$$

$$C^2\left[X\right]=\frac{V\left[X\right]}{E^2\left[X\right]},\ j=0,1,2,\ldots \quad (5)$$

A Markov process is defined by its transition matrix P and the initial state

$$p=F_X\left(quantile\left(p\right)\right)=\int_{-\infty}^{quantile(p)} f\left(x\right)dx \text{ X continuous}.$$

Birth and Death Processes

A birth – death process is a class of Markov chains where time happens continuously (Taylor and Karlin, 1984). In these types of processes there is a set of states {0, 1, 2 …} that is called "population" and the transitions between states indicate increments or decrements in the number of individuals in the "population".

When the system is found in a state $n \geq 0$, a period of time elapses before the arrival of an individual that generates the transition to a state with $n+1$individuals; time is a random variable with exponential distribution and a rate λ_n. This process is known as "birth" precisely because it involves increasing the number of individuals present in the system.

On the other hand, when the system is in a state $n \geq 0$, a period of time elapses before an individual leaves generating the transition to a state with $n-1$individuals; time is also a random variable with an exponential distribution and rate μ_n. This process is known as "death" because the population decreases (Figure 2).

Figure 2. Birth-Death process with rate of birth λ and death rate μ

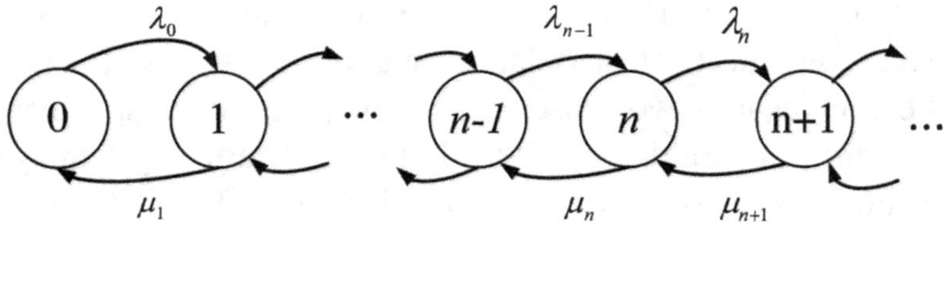

From the model of the "birth" and "death" process, we calculate the probability p_n of the system being in state n, in other words the probability of there being n individuals. For this purpose, a flow balance model must be developed for the entry and exit of a state: for each state, the speed of entry to said state must be equal to the speed of exit from said state.

M/M/1 System

If the inter-arrival time and service time follow an exponential distribution, then this is a particular case of Markov chains. We will now develop a model to get the probability of finding the system in a certain state and then use this result to get the corresponding equation for the calculation of the amount of work in process of an M/M/1 system.

Let us assume that the customers arrive at a single-server queue with time intervals of $1/\lambda$ units of time per customer and the server provides the service at a speed of μ customers per unit of time. The scheduling policy is First In, First Out. Let us also assume that both the inter-arrival time and the service time are random variables and follow an exponential distribution. There is no constraint on the number of customers that can stay in the system, however, as we stated in chapter 1, in order to avoid the queue endlessly growing, it is essential for the service capacity μ to be greater than the number of customers λ that arrive at the system, in other words congestion must be $\rho = \lambda/(c\,\mu) < 1$, we assumed that there is only one server so $c=1$. Finally let us assume that the system has been operating for a long time, so we know it is in a stable state.

In queues, "births" are generated by the arrival of customers in the system and "deaths" happen when a customer finishes its service and leaves the system. States in this Markov chain correspond to the total number of customers in the system at any given time (including the ones in the queue and the one being served).

The transition from state n to state $n+1$ happens at a rate of λ times the probability p_n of remaining in said state, in other words λp_n. Furthermore we have the output of customers whose service has finished which lowers the number of customers in the system and the transition from state n to state $n-1$ happens at a rate of μ times the probability p_n of residences in said state, in other words μp_n. Both flows are equal in stable state (Figure 3).

In order to construct the flow balance equations we will use a technique that first cuts the nodes into two subsets and then applies the respective balance equations (Curry and Feldman, 2009). Let us look at the system again. The first cut is made between state (node) 0 and state (node) 1 (Figure 4).

Figure 3. Number of clients in an M/M/1 queue with arrival rate λ and service rate μ

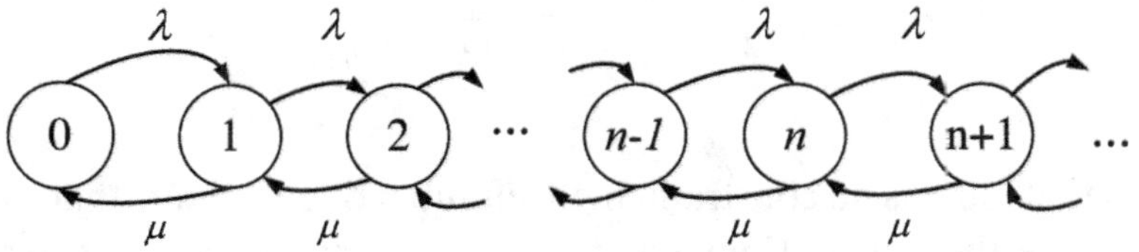

Figure 4. First cut

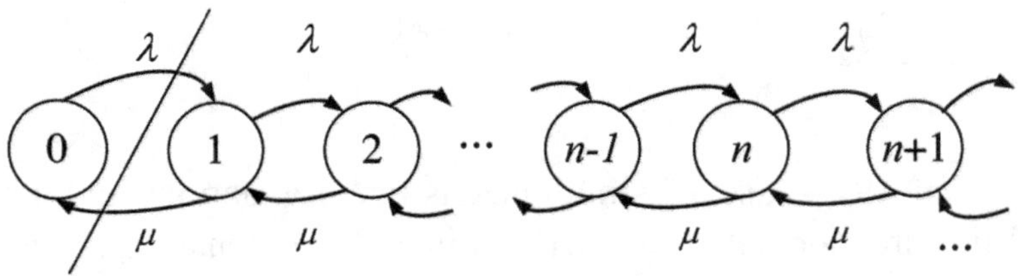

The respective flow balance equation is:

$$f\left(x\right) = \frac{1}{b-a} a \leq x \leq b \tag{6}$$

The second cut is made between node 1 and node 2 (Figure 5).
The flow equation is:

$$F\left(x\right) = P\left(X \leq x\right) = \frac{x-a}{b-a} a \leq x \leq b \tag{7}$$

We shall make an additional cut for the next state. The respective balance equation is:

Figure 5. Second cut

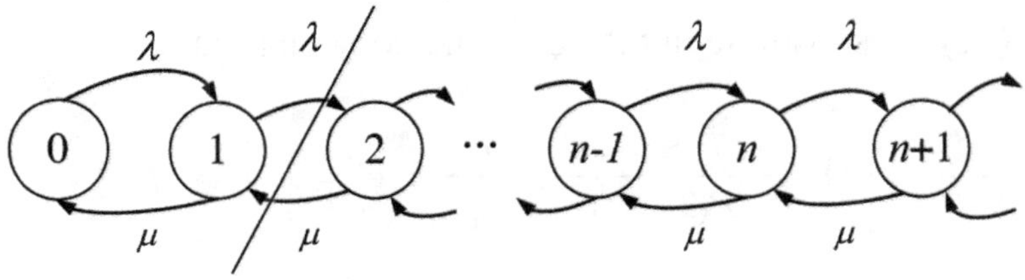

$$E\left[X\right]=\frac{a+b}{2};V\left[X\right]=\frac{\left(b-a\right)^{2}}{12};C\left[X\right]=\frac{b-a}{\sqrt{3}\left(a+b\right)} \tag{8}$$

Given that there is no constraint in the number of customers that can reside in the system, there is no final equation. The resulting system is as follows:

$$f\left(k\right)=\frac{n!}{k!\left(n-k\right)!}\left(p\right)^{k}\left(1-p\right)^{n-k}=\frac{n!}{k!\left(n-k\right)!}\left(\frac{\lambda}{n}\right)^{k}\left(1-\frac{\lambda}{n}\right)^{n-k}\quad \text{for } k=0,1,\cdots. \tag{9}$$

The system (9) is a set of simultaneous linear equations and, given that probabilities are calculated, we need to include an equation to ensure that the sum of the probabilities is 1. This is known as a normalization equation: $p_0 + p_1 + \ldots + p_n = 1$; or in condensed form $\sum^{\infty}{}_{n=1} = 1$. Then the system is as follows:

$$E\left[N\right]=np=\lambda;V\left[N\right]=np\left(1-p\right)=npq;C^{2}\left[N\right]=\frac{1-p}{np}=\frac{q}{np} \tag{10}$$

Each probability p_n shall now be expressed as a function of p_0:

$$\begin{aligned}f\left(k\right)&=p\left(1-p\right)^{k-1}=pq^{k-1}\\F\left(k\right)&=1-\left(1-p\right)^{k}=1-q^{k}\end{aligned}\quad \text{for } k=1,2 \tag{11}$$

By generalizing, we get:

$$E\left[N\right]=\frac{1}{p};V\left[N\right]=\frac{1-p}{p^{2}};C^{2}\left[N\right]=1-p \quad , \qquad \text{f o r}$$

$$P\left(T>k\right)=1-P\left(T\leq k\right)=1-F\left(k\right)=1-\left(1-q^{k}\right)=q^{k} \tag{12}$$

Now (12) we substitute in the normalization equation:

$$P\left(T>t+s\middle|T>t\right)=\frac{P(T>t+s)}{P(T>t)}=\frac{q^{t+s}}{q^{t}}=q^{s}=P(T>s) \tag{13}$$

Rearranging this expression to get p_0:

$$k = \frac{\log\left(rand\right)}{\log\left(1-p\right)} \tag{14}$$

$$\lambda = np \tag{15}$$

A geometric series like the denominator one has the following property:

$$\lambda = np \tag{16a}$$

By applying this property to the result obtained we are able to rearrange the expression as follows:

$$f\left(k\right) = \lim_{n\to\infty}\left[\frac{n!}{k!\left(n-k\right)!}\left(\frac{\lambda}{n}\right)^{k}\left(1-\frac{\lambda}{n}\right)^{n-k}\right] = \frac{\lambda^{k}e^{-\lambda}}{k!} \text{ for } k = 0,1,\cdots. \tag{16b}$$

Or alternatively, as a function of the congestion of the system:

$$E\left[k\right] = \lambda \tag{17}$$

As we can see, the server must have sufficient capacity to meet the demand, if the condition $\rho < 1$fails then the results becomes meaningless, mathematically speaking.

When the condition is fulfilled, then, as the service rate gets closer to the value of the arrival rate, the probability of finding the system empty gets increasingly smaller and, conversely, the more capacity the server has, the greater the probability of finding the system empty.

Performance Measures

Using the resulting probability expressions, we construct the equations for the average amount of work in process and the cycle time in system.

Average Work in Process

Recovering the equation (5) of WIP_S from Table 2 in chapter 1 and by substituting the probability expressions obtained in the above section, we get:

$$V\left[k\right] = \lambda \tag{18}$$

The summation can be simplified by applying the following property:

$$C\left[k\right] = \frac{1}{\sqrt{\lambda}} \tag{19a}$$

Rearranging (18):

$$\lambda_L = \frac{1}{2n_{2K,\alpha/2}}^2$$
$$\lambda_U = \frac{1}{2n_{2K+2,1-\alpha/2}}^2 \tag{19b}$$

Because the station has a single server, so the work in queue is

$$^2_{df,prob} \tag{20a}$$

By substituting the probability expressions, we arrive at:

$$P\left(k=0\right) = \frac{\lambda t^0 e^{-\lambda t}}{0!} = e^{-\lambda t} \tag{20b}$$

If the service rate is very close to the arrival rate, then the number of customers in the system grows rapidly, in other words, we need enough capacity in a queueing system to serve the customers that arrive in the system, otherwise the number of customers waiting will rise very fast (Figure 6).

Figure 6. Work in process in the system vs congestion

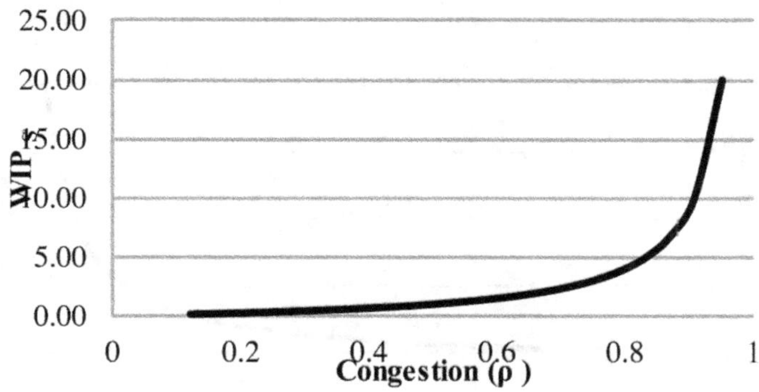

Average Cycle Time

As has been mentioned, this is the time that an entity remains in the system: from its entry, lining up waiting to be served, receiving its service and leaving the place. To get an equation we will use Little's Law, $WIP_S = \lambda CT_S$. The time spent in the system is:

$$P(T \le t) = 1 - e^{-\lambda t} = F(t) \text{ for } t > 0 \; f(t) = \lambda e^{-\lambda t} \text{ for } t > 0 \tag{21}$$

Likewise, the time in queue has the following expression:

$$\begin{aligned} E[T] &= \frac{1}{\lambda} \\ V[T] &= \frac{1}{\lambda^2} \\ C[T] &= 1 \end{aligned} \tag{22}$$

It is essential for the capacity to be higher than the service demand; otherwise the cycle time becomes infinitely long (Figure 7).

Example 3.1: Let us assume that a cafeteria has only one person providing the service: they take the order, prepare the coffee and charge. The average service time is 7 minutes per order or, in capacity terms, they attend 9 customers an hour. The arrivals are at a rate of 8 customers an hour. Assume that the service time and the time between arrivals are distributed exponentially. The performance measures of this system are:

Figure 7. Cycle time vs. congestion (ρ)

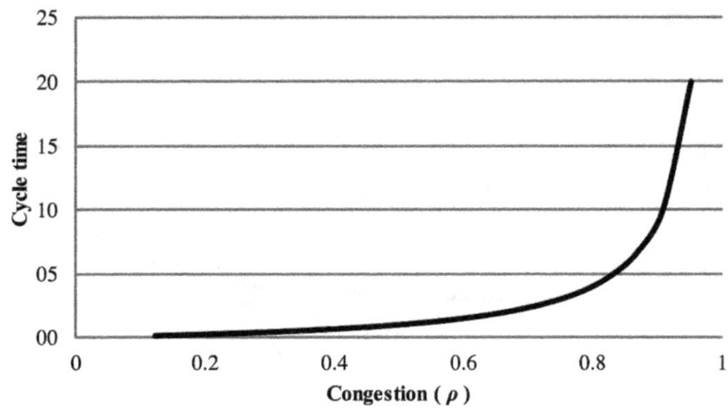

- **Congestion:** $\rho = 8/9 = 0.88$
- **From Expression (21), Average Cycle Time in System is:** $CT_S = 1/(9-8) = 1$ hour
- **From Expression (22), Average Cycle Time in Queue:** $CT_q = 0.88/(9-8) = 0.88$ hour
- **The Average Work in Process in the System is Obtained With (19b):** $WIP_S = 0.88/(1-0.88) = 7.33$ customers
- **The Average Work in Process in the Queue From Expression (20b) is:** $WIP_S = (0.88)^2/(1-0.88) = 6.45$ customers

In queues, not only the work in process and cycle time should be considered as the sole performance measures, we also need to analyze their implications in greater depth, as in the following part of the analysis. For example, the time spent in the queue represents 100 (0.88/1) = 88% of the total time in the system, carefully observing that the customer has to wait 0.88 hours (53 minutes) while the service time takes about 17 minutes. (In systems where waiting is critical, such as a hospital where the administrators need to find strategies to ensure that a customer does not wait too long in the queue).

Continuing on with the earlier example, there are some $6.45 \approx 7$ customers. It is common for a customer to reject the idea of a long wait.

The following sections will show how to estimate the capacity of a queueing system that is required for it to meet its objectives. This is the point where mathematical tools make it possible for us to visualize the state of the system (descriptive stage) and then they are later applied to analyzing the effect of changes (prescriptive stage).

M/M/C System

In real life, parallel-server systems are much more common. In order to find the equations for the system's performance measures, we once more assume that the time between consecutive customers is a random variable and follows an exponential distribution. The system now has c servers that operate independently, while the service time is also a random variable and follows an exponential distribution. Let us also assume that there is no difference between servers.

Adding servers is equivalent to increasing the system's service capacity, in other words, it will now have a service capacity of $c\mu$ customers per unit of time. In a system with a variety of servers, the system's capacity or service rate will depend on the number of customers in the system, because the servers could be fully occupied or there may be servers that are free or vacant at any specific moment. In the one case, there are fewer customers in a particular instant than servers $n < c$, therefore the number of free servers is $c - n$ and the service rate will be $n\mu$; whereas in the other case, the number of customers is equal to or more than the number of servers, therefore the service capacity will be $c\mu$.

The service rate when there are n customers shall be denoted as μ_n, and is written as follows:

$$P\left(T > t + s \middle| T > t\right) = \frac{P(T > t + s)}{P(T > t)} = \frac{e^{-\lambda(t+s)}}{e^{-\lambda t}} = q^s = P(T > s) \tag{23}$$

It is worth mentioning that demand does not depend on the number of customers in the system, therefore $\lambda_n = \lambda$. Using the definition from the birth-and-death process, the steady-state probabilities p_n for the system are:

$$t = -\frac{\log\left(rand\right)}{\lambda} \tag{24}$$

For example the probability of the system being empty has the following expression:

$$\overline{t} = \frac{\sum_{i=1}^{n} t_i}{n} \tag{25}$$

Again, if $\rho > 1$ then the results are meaningless.

Performance Measures

From the equations for states n, we get the expressions for the system's work in process and cycle time.

Average Work in Process

The work in queue is obtained by substituting the flow equations; the final expression is as follows:

$$\bar{t} \tag{26}$$

The work in system is obtained by substituting in the general formula for the work in system:

$$f(t) = \lambda \frac{(\lambda t)^{k-1}}{(k-1)!} e^{-\lambda t} \text{ for } t > 0 \tag{27}$$

Average Cycle Time

The time in system is obtained by using Little's Law:

$$F(t) = 1 - \sum_{j=0}^{k-1} \frac{(\lambda t)^{j}}{j!} e^{-\lambda t} \text{ for } t \geq 0 \tag{28}$$

Finally, the time in queue is:

$$1/\sqrt{k} \tag{29}$$

Example 3.2: A terminal services several passenger bus lines. There are 4 spaces in the terminal for the buses that arrive at the station. While a vehicle is parked, a routine inspection is performed, the luggage is put into the compartment and the passengers board the vehicle. This operation takes around 15 minutes and is exponentially distributed. 12 buses arrive at the terminal every hour.

This terminal's performance measures are as follows: This is where a spreadsheet will allow us to speed up the operations.

The congestion (or fraction of capacity used) is 0.75 or in other words, the system is found to be at ¾ of its rated capacity.

The time in system (equation 28) is 0.377 hours or 22.64 minutes, of which the bus spends 7.64 minutes waiting for a space to be vacated, in other words, waiting in the queue represents 33.75% of the total time.

On average there would be 4.52 buses in the terminal (equation 27) and the average queue has around 1.52 buses (equation 26).

What is the probability of finding the 4 spaces occupied? The probability of finding $n = 4$ buses should then be calculated and, based on equation 24, the result is $p_4 = 0.58$

Two additional spaces are set up. How do the long-term performance measures change when these new spaces enter into operation? The original performance measures are contrasted with the new scenario in Table 2.

In general, when servers are added, there is a drop in the amount of work in process and cycle time. It must be remembered that the capacity increases with each server that is incorporated into the system, which is the reason for such pronounced drops.

Before implementing the proposed changes, it is also advisable to carry out an analysis that incorporates the economic aspect. Adding servers is beneficial because the queue is shortened but also implies that servers will more often be left idle. In example 3.2, the capacity used with the additional servers is 50%. What happens to the rest of the capacity? The answer is that the 6 spaces lie idle half the time, which would be questionable in economic terms. This is the downside: wasted capacity.

Further on we will go into the issue of adding servers to increase the capacity of a queue in more depth.

Table 2. Performance measures at c=4 and c=6

Measure	4	6
ρ	0.75	0.5
p_0	0.03	0.048
WIP_q	1.52	0.099
WIP_s	4.52	3.099
CT_q	0.127 hours (7.64 minutes)	0.00826 hours (0.49 minutes)
CT_s	0.377 hours (22.64 minutes)	0.258 hours (15.49 minutes)

M/M/1/K System

In the two models reviewed in the earlier sections, we assumed that the system has enough spaces to hold any number of customers that arrive in the system. As has also been mentioned earlier, in M/M/1 and M/M/c systems, the more the service demand approaches the value of the capacity ($\lambda \rightarrow c\mu$), the more the amount of work in process WIP (not only of the system but also in the queue) grows very fast. It is also assumed that all the customers that arrive, line up and are served.

A situation where the server does not have enough capacity in the operation to serve the customers implies chaos: one typical example is when there is a long-awaited movie premiere, the lines to buy tickets are very long and it is normal for such lines to be several blocks long.

In factories and some systems however, the number of available spaces (K) is usually not enough to hold any amount of pieces waiting to be processed. Among other things, it is good manufacturing practice to keep the work area in good order and for this it is critically important to keep the work in process under control.

Let us assume that the customers arrive at a single-server queue with time intervals of $1/\lambda$ units of time per customer and the server provides the service at a rate of μ customers per unit of time. Let us assume that both the time between customers and the service time are random variables and follow an exponential distribution. The scheduling policy is First In, First Out. Finally let us assume that the system has operated for a long time, thus ensuring that it is in a stable state.

Let us assume that unlike the M/M/1 and M/M/c systems, the space for the customers is now limited to a certain number of spaces K (1 receiving service and the rest in a queue). When there are places or spaces available, then any customer that arrives at the system shall be able to enter and join a queue; however, if all the spaces are occupied, then any customer that arrives at that time shall be rejected and lost. As the maximum number of customers that can be in the system is K, then it is not necessary for the capacity to be more than the service demand or, in other words, $\rho < 1$ is not necessary (Figure 8).

The analysis to get the transition probabilities is similar to the one used for the M/M/1 system but with the distinction that now the possible states are $n = 0, 1, 2, \ldots, K$ (Figure 9).

The respective balance equations, together with the normalization equation, are:

Figure 8. Queue system with limited capacity for the work in process

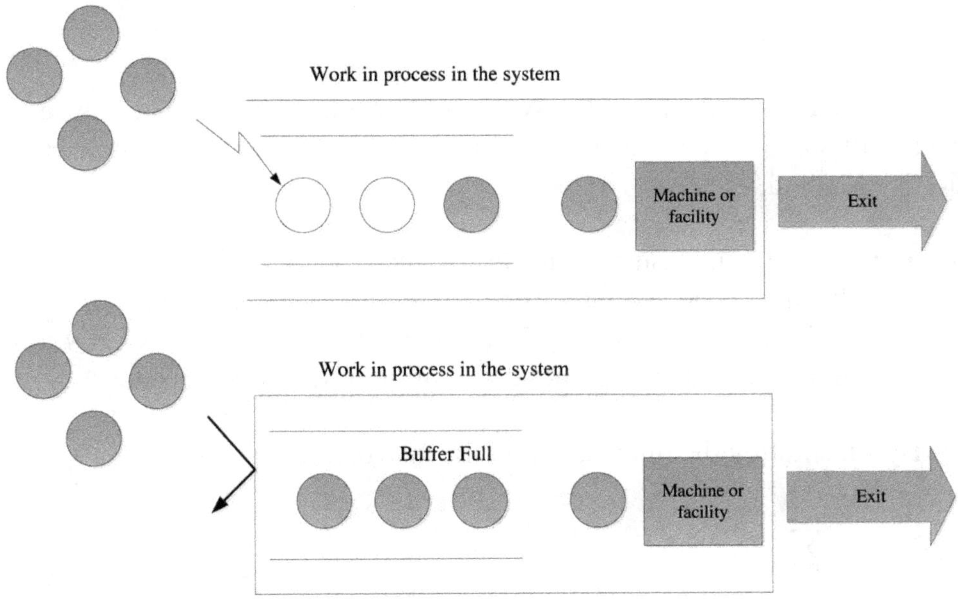

Figure 9. Number of clients in an M/M/1/K queue with arrival rate λ, service rate μ and capacity K

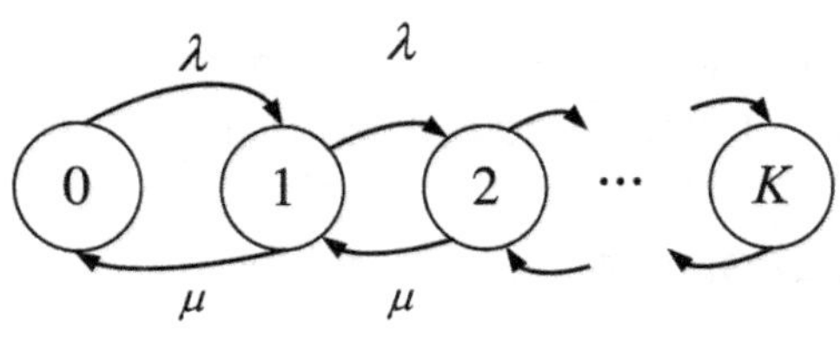

$$t = -\frac{1}{\lambda}\log\left[\prod_{i=1}^{k} rand_i\right] \tag{30}$$

Rearranging the above system we get:

$$\bar{t} \tag{31}$$

The above system can be represented in matrix form as follows:

$$\bar{t} = \frac{\sum_{i=1}^{n} t_i}{n}$$

If λ and μ are known then the system of linear equations can be solved by means of some of the known procedures, such as, for example, the Gauss-Jordan Method.

Example 3.3: Let us assume that a factory has a work station with space for 3 jobs in a queue and the machine only processes one job at a time ($K = 4$); the capacity is $\mu = 5$ jobs a minute and the demand is $\lambda = 4$ jobs a minute. Get the probability distribution of the possible states.

By substituting the information in the matrix, the system is:

$$E\left[\bar{t}\right] = E\left[\frac{\sum_{i=1}^{n} t_i}{n}\right] = \frac{1}{n} E\left[\sum_{i=1}^{n} t_i\right] = \frac{1}{n}\left(\frac{n}{\lambda}\right) = \frac{1}{\lambda} = E\left[T\right] \tag{32}$$

The solution of a system of linear equations is x= A^{-1}b. Using the data for the problem, the probabilities are as shown in Table 3.

The probability of finding the system empty is 0.297. The probability of a customer being rejected because of lack of space is 0.121 (Table 3).

Performance Measures

While there is at least one space, the customers shall continue to enter the system and line up to wait their turn. When all the spaces are occupied, in other words, when the number of customers reaches the value K, then "the

Table 3. Probability of finding n customers in the system

State	p_n
0	0.297
1	0.237
2	0.190
3	0.152
4	0.121

door is closed" and any customer that tries to enter will be rejected. The number of customers that enter the system is represented as follows:

$$V\left[\bar{t}\right]=V\left[\frac{\sum_{i=1}^{n} t_i}{n}\right]=\frac{1}{n^2}V\left[\sum_{i=1}^{n} t_i\right]=\frac{1}{n^2}\left(\frac{n}{\lambda^2}\right)=\frac{1}{n}\left(\frac{1}{\lambda^2}\right)=\frac{1}{n}V\left[T\right] \quad (33)$$

We must point out the following:

1. Although the demand is greater than the service capacity, the queue will not grow endlessly because, if the spaces are fully occupied, the door stays "closed" and any customer that arrives will not be able to line up.
2. Although customers are not permitted to enter the system, it is still operating and providing the service.

Probability of Finding N Customers in the System

Based on aforementioned subparagraphs, it must be pointed out that it is not necessary for the condition $\rho < 1$ to be fulfilled in M/M/1/K systems. From the flow equations we get the expression for calculating the probability of finding n customers in the system:

$$\bar{t} \quad (34)$$

In particular, the probability of the system being empty is:

$$E\left[\bar{t}\right]=E\left[T\right] \quad (35)$$

The probability of all the spaces in the system being occupied ($n = K$) when $\rho < 1$ is:

$$V\left[\bar{t}\right]=\frac{1}{n}V\left[T\right] \quad (36)$$

Throughput

One of the characteristics of this system that has been remarked upon is that when all the available places are occupied, customers are rejected and lost; by rejecting customers, the number of served customers that leave the system through "the exit door" will be less than the original demand that arrives through "the entrance door". In M/M/1 and M/M/c systems, the number of customers that enter is equal to the number of customers that leave, in other words $\lambda = \lambda_{out}$, whereas for the M/M/1/K systems: $\lambda > \lambda_{out}$.

In the fraction of time that the system permits entry $1 - p_K$, those customers for which there is space in the system are served; then the expression for the customers that were served is:

$$\lambda\overline{x} \tag{37}$$

Average Work in Process

The amount of work in process in the system is obtained from the expression (20a):

$$P\left(E_{\text{inf}} \leq \lambda\overline{x} \leq E_{\text{sup}}\right) = P\left(E_{\text{inf}} / \overline{x} \leq \lambda \leq E_{\text{sup}} / \overline{x}\right) = 1 - \alpha \tag{38}$$

For the size of the queue the general expression is $WIP_q = WIP_S - \lambda / \mu$ and from this we get the following formula:

$$f\left(x\right) = \frac{1}{\sqrt{2\pi\sigma}} e^{-\left(x-\mu\right)^2/\left(2\sigma^2\right)} \tag{39}$$

Average Cycle Time

The time a customer stays in the system only applies to customers that have been served and is obtained by using Little's Law:

$$z = \frac{x - \mu}{\sigma} \tag{40}$$

Finally the time in queue:

$$\left(z\right) = \frac{1}{\sqrt{2\pi}} e^{-\left(z\right)^2/2} \tag{41}$$

Example 3.4: Let us consider an item of equipment that processes pieces at the rate of 20 an hour. At the present time a "buffer" is considered to keep material ready to be fed into the equipment and avoid an excessive accumulation of material. The place used for processing the pieces has capacity for 10 units. The demand of pieces is 18 an hour. Assume that both service time and time between arrivals are exponentially distributed.

The operations manager wants to know: 1. The probability of the available space in the machine being full. 2. The available amount of work in process in the machine (including the piece being processed) and the time spent in the system. 3. The machine's real production rate (throughput). 4. What happens with the number of customers leaving, if the space is successively reduced to 5 spaces?

1. From the equation (34), the probability p_K of the system being full is 0.0508.
2. The average amount of work in process (equation 38) is 3.969 pieces. The time in system (equation 40) is 0.23 hours or 13.93 minutes $\approx$ 14 minutes; the time in a queue (equation 41) is 0.182 hours or 10.93 minutes $\approx$ 11 minutes.

The fraction of time in a queue is 10.93/ 13.93 = 0.7847 or approximately 78.47% of the total residence time in the system.

3. The effective exit is 17.08 customers (equation 37). In the long term 18- 17.08 = 0.92 customers $\approx$ 1 customer are lost on average (do not enter the system because it is blocked).
4. On reducing the available space, the output (th) from the system decreases and in consequence the system stays full most of the time, ending up with a higher number of customers being rejected. When K= 5 on average approximately 2.23 $\approx$ 3 customers are lost. Likewise the time in system decreases, when K=5, the residence time is 0.1395 hours or around 8.37 minutes. Finally the drop in production is 7.9%.

One positive effect of limiting the work in process is that the cycle time is shortened, because the average size of the queue is limited. However the result tells us that the decision to limit the space available in a queue has the negative effect of reducing the customers leaving the system.

One common practice in production lines is to limit the amount of work in process by restricting the size of the queue (the so-called buffer) however, as can be appreciated, this decision must not be taken lightly; not all factories or service companies can give themselves the luxury of producing 8% less.

Finally it must be pointed out that if the number of spaces is increased, in other words, if the value K is made very large ($K \to \infty$), the system shall behave like an M/M/1 queue.

M/M/C/K System

The single-server case seen in the above section is a particular instance of the M/M/c/K system. The arrival rate and the rate service μ_n have the following expressions:

$$\overline{x} \tag{42}$$

$$\overline{x} = \frac{\sum_{i=1}^{n} x_i}{n} \tag{43}$$

The respective probability expressions p_n:

$$z = \frac{\overline{x} - \mu}{\sigma / \sqrt{n}} \tag{44}$$

When therefore $n = K$ is the probability of the system being full. Moreover the probability of finding the system empty is:

$$\overline{x} \tag{45}$$

Performance Measures

The respective expressions for the work in process and cycle time are given below.

Average Work in Process

In the $\rho = 1$ case, the work in queue is:

$$\overline{x} \quad (46)$$

Finally for the $\begin{matrix} z_1 = \cos\left(2\pi * rand_1\right)\sqrt{-2\log\left(rand_2\right)} \\ z_2 = \sin\left(2\pi * rand_1\right)\sqrt{-2\log\left(rand_2\right)} \end{matrix}$ case we have that:

$$x = \mu + \sigma z \quad (47)$$

In the case of work in process of the system, only those customers that entered the system during the periods when there was space should be considered:

$$\overline{\mathrm{x}} \quad (48)$$

The cycle time is obtained by applying Little's Law.

The code given below is written in Scilab and calculates the properties of M/M/c/K systems.

```
function [x]=mmck(mu,lambda,c,K)
ro=lambda/(c*mu)
// Probability of the System being empty
if (ro==1) then
cc=c-1
sum=0.0
roo=lambda/mu
for j=0:cc
cte=roo^j/factorial(j)
sum=sum+cte
end
p0=roo^c/factorial(c)
p0=p0*(K-c+1)
p0=sum+p0
p0=1/p0
else
cc=c-1
sum=0.0
```

```
roo=lambda/mu
for j=0:cc
cte=roo^j/factorial(j)
sum=sum+cte
end
p0=(1-ro^(K-c+1))/(1-ro)
p0=p0*roo^c/factorial(c)
p0=sum+p0
p0=1/p0
//p0=sum
end
//Probability distribution for the number of customers in the
system
p(1)=p0
i=2
p(1)=p0
while (i<=K+1)
if(i-1<c) then
p(i)=p0*roo^(i-1)/factorial(i-1)
else
p(i)=p0*roo^(i-1)/(c^((i-1)-c)*factorial(c))
end
i=i+1
end
[x]=[p]
//Performance measures
if ro==1 then
wipq=roo^c*(K-c)*(K-c+1)/(2*factorial(c))
wipq=p0*wipq
else
wipq=roo^(c+1)/((c-roo)^2*factorial(c-1))
wipq=wipq*(1-(ro^(K-c+1))-(K-c+1)*(1-ro)*ro^(K-c))
wipq=wipq*p0
end
wip=wipq+lambda*(1-p(K+1))/mu
tc=wip/(lambda*(1-p(K+1)))
leff=lambda*(1-p(K+1))
[x]=[ro,p0,p,wipq,wip,tc,leff]
endfunction
```

Example 3.5: Let us reconsider the system described in example 3.5. Now we want to increase the capacity of the station by adding a parallel machine, so the system shall now be M/M/2/K.

The results obtained using this program are $WIP_S = 1.126$, $WIP_q = 0.2264$, cycle time = 0.0625 hours and Th = 17.99.

Rules for Determining the Number of Servers

It is often necessary to increase the capacity of a system to improve its performance; the increased capacity is achieved either by altering the speed of the server or by adding parallel servers to serve the customers.

The criterion for determining how many additional servers are needed can be varied according to the situation: For some administrators it is enough to ensure that the system has the minimum amount needed to meet the service demand, while others may prefer to set a target (such as the probability of waiting) which may lead to a different result. The approaches for determining the number of servers are (Gross, Shortle, & Thompson, 2008):

1. **Based on the Quality of the Service:** A level of congestion ρ must be set, then servers are added until the desired level of congestion is obtained.
2. **Based on the Efficiency of the System:** Enough servers are added for the system to be stable, in other words that $\rho < 1$.
3. **Equilibrium Between the System's Efficiency and the Quality of the Service:** A metric connected with the level of congestion ρ is selected, for example, the probability of a customer having to wait. Then a target is set for this parameter and, finally, servers are added until the target is achieved.

The result can be very different depending on the approach.

Example 3.6: A health center has 4 doctors to look after the people who arrive seeking some type of attention. The time spent attending each patient requires a mean of 20 minutes and is exponentially distributed. The service demand currently exceeds the capacity of the doctors and a sampling indicates that, on average, 15 patients arrive to the health center every hour, the time between arrivals is exponentially distributed. How many doctors are needed to meet the demand?

1. **Focusing on the Quality of the Service:** Let us assume that it is decided to work with a level of congestion of under 0.8 (80% of the capacity). The search is given in Table 4, with the answer being the addition of 3 doctors, thus ending up with a staff of 7:
2. **Focusing on the Efficiency of the System:** In this case, we only need to add 2 doctors to the staff in order to have enough capacity, so there will 6 doctors in the office.

Table 4. Congestion value (ρ) with c doctors

Servers	Current 4	5	6	7	8
ρ	Unstable	Unstable	0.833	0.7143	0.6250

3. **Seeking to Have a Balance Between Efficiency and Quality:** The probability of a customer having to wait is an example of a congestion-related measurement (Gross, Shortle, & Thompson, 2008). Let us assume that we want the probability of a patient finding all the doctors occupied (and, in consequence, having to wait) to be no more than 0.4. This calculation can be done using the following formula (Gans, Koole, Mandelbaum, 2003):

$$\overline{x} = \frac{\sum_{i=1}^{n} x_i}{n}$$
$$S^2 = \frac{\sum_{i=1}^{n} \left(x_i - \overline{x}\right)^2}{n-1} = \frac{\sum_{i=1}^{n} \left(x_i\right)^2 - n\overline{x}^2}{n-1} \qquad (49)$$

where (49) is the probability of finding *n* servers occupied; term $r = \lambda / \mu$ is the workload of the system; the equation is known as the Erlang-C formula. Another option is to calculate the cumulative probability of having *n* customers in the system, using *n* servers, and get the result from the difference.

The workload is $r = 15 / 3$. The search is started with 6 doctors as 4 and 5 doctors do not provide enough capacity. If there is a staff of 6 doctors, the cumulative probability of finding 5 doctors occupied is 0.4125(equation 48), then the probability of finding the 6th doctor occupied is 1- 0.4125 = 0.5875. The reasoning is the same as in the other two cases.

The answer then is to have a staff of 7 doctors (3 doctors in addition to the original staff). The probability of a patient having to wait (with the current demand) will be 0.3241, so we shall not exceed the established target of 0.4 (Table 5).

The result can be found by using the probabilities of finding *n* patients in the system. Table 6 shows the cumulative probabilities of finding *n* patients being attended by *n* doctors.

Table 5. Comparison of results for 6, 7 and 8 doctors

Staff	6 doctors	7 doctors	8 doctors
Probability of Having to Wait (All the Doctors Being Occupied)	1 – 0.4125 = 0.5875	1- 0.6759 = 0.3241	1-0.8327 = 0.1673

Table 6. Probability of finding n doctors occupied

Doctors in Consultation	Staff of		
	6 Doctors	**7 Doctors**	**8 Doctors**
	Cumulative P		
0	0.0045	0.0060	0.0065
1	0.0271	0.0358	0.0388
2	0.0835	0.1105	0.1198
3	0.1775	0.2350	0.2547
4	0.2950	0.3906	0.4233
5	**0.4125**	0.5462	0.5919
6	Free doctor	**0.6759**	0.7324
7		Free doctor	**0.8327**
8			Free doctor

In the case where there are 6 doctors, the probability of finding 5 people being seen by a doctor at the same time is 0.4125, while the probability of finding the 6th doctor occupied is 1 – 0.4125 = 0.5785.

These rules allow us to analyze any decision on the basis of the performance of the system and the administrator just needs to set the desired performance level. We must add another option for estimating the number of servers, which is the following equation:

$$\overline{x} \tag{49}$$

(49) is known as the square root rule for the number of servers. The second term is the number of extra servers for a work load r, $0 < \beta < \infty$ and is known as the service level. Borst et al (2000) developed the following approximation for values of r when estimating the optimum value of parameter β (Borst et al 2000, Gans, Koole, Mandelbaum, 2003):

$$\overline{x} \tag{50}$$

Expression (50) can be applied in the case where service costs and waiting costs are linear in respect of time.

Example 3.7: Equation (50) shall be applied to answer the question with approach number 3. With the data $r = 15 / 3 = 5$, then $\beta = 1.485$; the number of additional servers is $1.485\sqrt{5} = 3.32 \approx 3$.

The total number of doctors must be $c = 5 + 3.32 = 8.32 \approx 8$.

The following section also gives examples of decision-making when taking the economic aspect into account.

Costs in Queues

In the examples posed earlier the models have been used for purely descriptive purposes, in other words they only give information about the performance measures for certain operating conditions. However these can be used (and in fact are designed for that purpose) for situations where we want to control the system or for cases where the design of the system is proposed or required. The advantage of an analytical model is that it shows how every variable relates to the others. In the case of queues, this is of interest when considering the behavior of a system:

1. When the demand changes.
2. When there are changes in the service time.
3. When new servers are added.

The above are only some of the alternatives. The objective of the study is generally to answer a question of the "What would happen if…?" type

It is worth mentioning that decisions often imply the aspect of the cost. Changing the conditions of a station (be they machines or people) implies some type of monetary investment: changes in equipment, staff training, changes in the job and adding a "buffer" are just some examples. In this sense, it is important to take the relevant costs into account and quantify the effects on a monetary basis.

Cost Model

In a queueing system, the total expected operating cost per unit of time is the sum of the expected cost of the server plus the expected cost of the customers that are in the system:

$$\frac{\overline{x} - \mu}{s / \sqrt{n}} \tag{51}$$

In the case of a term corresponding to the service cost, the variables that interest us are the number of servers c and/or service time $1/\mu$ in the event that altering the speed of attention is a viable option. The cost of serving a person is $C_{service}$ and its units are \$/customer. By multiplying this constant by the total service rate of the station and the number of servers, the service cost function is:

$$P\left(t_{\alpha/2;n-1} < \frac{\overline{x} - \mu}{s / \sqrt{n}} < t_{1-\alpha/2;n-1}\right) = 1 - \alpha \tag{52}$$

If the waiting cost is proportional to the waiting time, then the function is as follows:

$$P\left(x - \mid t_{\alpha/2;n-1} \mid \frac{s}{\sqrt{n}} < \mu \left(\overline{x} + \mid t_{\alpha/2;n-1} \right| \frac{s}{\sqrt{n}}\right) = 1 - \alpha \tag{53}$$

where C_{Wait} is the cost attributed to each customer for every unit of time it remains in the system and WIP_S is the amount of work in system. The model is as follows (Figure 10):

$$t_{\alpha/2;n-1} \tag{54}$$

Using the cost model, an administrator can carry out a monetary analysis of the system based on the number of servers, the service rate or demand. On graphing the behavior of the cost as a function of the capacity we can see that, as the capacity of the system is increased, the cost decreases monotonously until it reaches a minimum value, then if the capacity continues to increase, the

Figure 10. Operation cost in a queue system

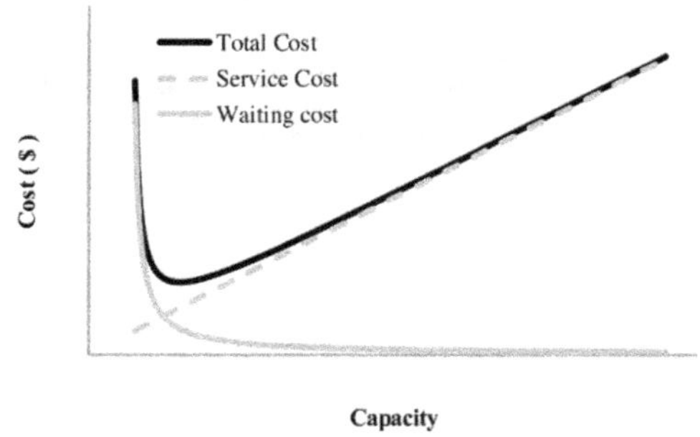

cost starts to rise again. The above therefore shows that finding an optimum capacity for the system is viable and this could be:

1. The optimum number of servers (c).
2. The optimum service rate (μ).

In case 1, a non-linear integer programming model must be solved as the number of servers is a variable integer, whereas in case 2, the service rate is a continuous variable and, as such, is a non-linear programming model.

Example 3.8: Let us consider example 3.2. As you will remember, there are 4 spaces in the terminal for the buses that arrive at the station. While the vehicle is parked, a routine inspection is performed, the luggage is put into the compartment and the passengers board the vehicle. This operation takes around 15 minutes and is exponentially distributed. 12 buses arrive at the terminal every hour.

The service given to each bus costs around $4 and includes the cost of the personnel who carry out the inspection, the person responsible for putting the luggage in the compartment as well as the cost of the personnel who check the passenger's ticket.

All the bus' seats are normally occupied. This represents about $60. For the purposes of the analysis, we shall consider that 10% of these revenues are the cost of waiting inside the terminal.

What is the terminal's operating cost under current operating conditions (4 spaces)?

A breakdown of the operating cost is given below. The service cost is:

$$\frac{(n-1)s^2}{\sigma^2}$$

Waiting cost

$$P\left({}^{2}_{\alpha/2,n-1} < \frac{(n-1)s^2}{\sigma^2} < {}^{2}_{1-\alpha/2,n-1} \right) = 1-\alpha$$

The total cost is: \$335 per hour. If the terminal operates 24 hours a day then the daily cost is \$8044.8.

How does the cost vary with 5, 6 and 7 spaces in operation? What number of servers would minimize the cost? Table 7 gives a breakdown of the respective costs. As can be seen, the total cost per hour of operation is \$335.2 under the current conditions. By successively increasing the number of spaces and assuming, moreover, that there are no changes in demand, the calculations indicate that it is a good idea to open up an additional space.

The result with 5 spaces lowers the total operating cost to \$281.24, whereas when increased to 6 spaces, the operating cost starts to rise again. From a cost perspective, it is convenient to operate with 5 spaces. This policy would ensure that any bus entering the terminal will find vacant spaces 60% of the time.

To complete the analysis, Table 8 shows the probabilities of finding *n* buses in the system with 4 and 5 servers.

In the instance with 4 spaces, the probability of finding the 4 spaces occupied is 0.127 (12.7% of the time), whereas working with 5 spaces, the probability of all of them being occupied is 0.094 (9.4% of the time).

Table 7. Breakdown of costs in each scenario

Item	4 Spaces	5 Spaces	6 Spaces
Service cost	64	80	128
Waiting cost	271.2	201.24	185.94
Total cost	335.2	281.24	313.94
ρ	75%	60%	50%
WIP_s	4.52	3.35	3.099

Table 8. Probability of finding n buses for c = 4 and c = 5 spaces

P_n	5 Spaces	4 Spaces
0	0.0466	0.037
1	0.139	0.113
2	0.209	0.169
3	0.209	0.169
4	0.157	0.127
5	0.0944	0.095

Example 3.9: Let us assume that we think it is viable to alter the time spent servicing the bus. What should the optimum service time be if we stay with just 4 spaces?

The results are given in Table 9. We start the search with a service rate of 3.5 buses / hour and increments of 0.5 units. We find that the optimum service rate is 7 buses an hour.

Example 3.10: Let us consider a system with a single server, with service time $1/\mu$ and time between customers $1/\lambda$, distributed exponentially. By substituting the expression (19) for the work in process of an M/M/1 system in the equation (54), we get an analytical expression of the cost for this type of system:

$$P\left(\frac{\left(n-1\right)s^2}{\underset{1-\alpha/2,n-1}{2}} < \sigma^2 < \frac{\left(n-1\right)s^2}{\underset{\alpha/2,n-1}{2}}\right) = 1-\alpha \tag{55}$$

Table 9. Total cost vs. service rate

Service Rate	3.5	4	4.5	5	5.5	6	6.5
Total Cost	514.4	335.2	277.2	249.8	234.4	226.2	221.6
Service Rate	**7**	7.5	8	8.5	9	9.5	10
Total Cost	**219.4**	219.6	220.6	222.7	225	224.9	290.2

If the decision variable is the service rate and assuming that it is a continuous variable, then the expression (55) can be derived and the criterion of the 1st and 2nd derivative applied to find an optimum μ. Then taking the 1st derivative:

$$ {}^{2}_{\alpha/2,n-1} \tag{56} $$

Equating to zero and solving for μ:

$$ A^2 = -n - \sum_{i=1}^{n} \frac{(2i-1)}{n} \left[\log F(x_i) + \log\left(1 - F(x_{n+1-i})\right) \right] \tag{57} $$

$$ (\mu - \lambda)^2 = C_{Wait} \frac{\lambda}{C_{Service}} \tag{58} $$

$$ \mu - \lambda = \sqrt{\frac{C_{Wait}\lambda}{C_{Service}}} \tag{59} $$

$$ \mu^* = \lambda + \sqrt{\frac{C_{Wait}\lambda}{C_{Service}}} \tag{60} $$

The second term of (60) is the additional capacity needed for the station to work at minimum cost and its behavior as a function of C_W / C_S is shown in Figure 11 with $\lambda = 1$.

Example 3.11: We will now, using a very simplified example, show the effect of limiting the (K) capacity of the system. Let us consider a piece of equipment that processes pieces at the rate of 20 pieces an hour. The station has space for 9 pieces, not including the one being processed. The pieces arrive at the rate of 18 pieces an hour. Let us assume that each finished piece is sold at \$30, the service cost is \$8, and the waiting cost is \$16.

Figure 11. Additional capacity (μ) as function of C_W / C_S and constant λ

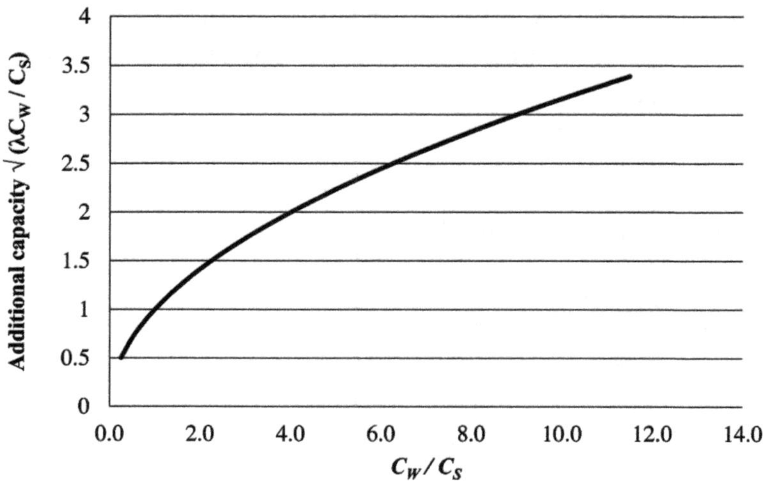

What is the operating cost and what is the total profit obtained?

The profit equation is

$$G = \text{Income} - \text{Costs} \tag{61}$$

$$G = I\lambda_{Out} - C_{Service}\mu - C_{Wait}WIP_S = 60\lambda - 8\mu - 16WIP_S \tag{62}$$

$$G = 60(16.929) - 8(20) - 16(3.646) = \$289.53 \tag{63}$$

How many spaces should be we keep, from the profit perspective?

We should remember that K controls the number of jobs present in the system (WIP_S). The rest of the parameters are constant. The search starts in K = 5 and the profit for each case is recorded in Table 10.

Table 10. Profit vs. K

K	5	6	7	8	9	10	11	12
Profit	276.831	283.67	287.4	289.1	289.5	289	288	286.4

By tabulating the profit results for each scenario we can see that it is advisable to keep a capacity of 9 places: 1 in the server and 8 in the queue; the profit will be \$289.53 (Table 9). Increasing this to more than 10 places in the system starts to lower the profit (Figure 12).

Exercises

1. Let us assume that students carry out a study of their university cafeteria, obtaining the following data: service time is 2 minutes per customer, the customers arrive at the rate of 1 every 2.5 minutes.
    a. Get the performance measures.
    b. What proportion does the time spent waiting in the queue represent?
2. Let us assume that for the above exercise the students recommend improving the service by having an extra person at the till. Construct a table comparing the congestion, the time in system, the work in process as well as the proportion of the waiting in the queue for $c = 1, 2, 3$ and 4 servers.
3. In a factory there is a work station that operates under the following conditions: the pieces arrive at the rate of 15 an hour. Every piece requires around 1/17 of hour. In front of the station there is only enough space for 3 pieces in the queue. Let us assume that the station only processes one piece at a time.

Figure 12. Profit vs. K

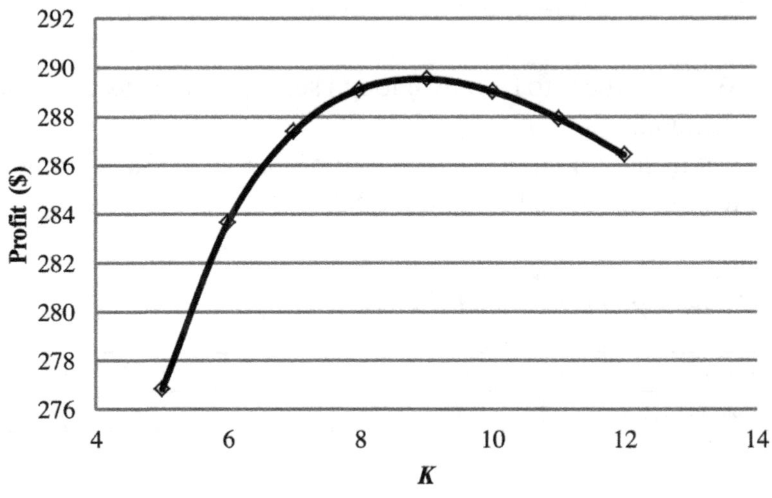

a. Get the time in system, the work in system, the probability of the system being blocked (p_k).
b. What is the productivity of the system (th)?
c. Let us assume that the management decides to fit out the area around the station to increase the space so that 7 pieces can be in the queue. How much does productivity increase in percentage terms? What is the new value of p_k? What is the new time in system?
d. Analyze the system (Work in process and time in system) for $K =$ 10, 100 and 1000. What conclusion do you reach about the behavior of the performance measures in comparison to the M/M/1 or M/M/c systems?
e. Let us assume that we have the following information: cost of the service $4.50, waiting cost per customer: $9, cost per lost customer: $12, revenues per piece sold: $ 50.
f. Get the value of K that maximizes the gain.
g. Get the number of servers that maximizes the gain.

4. In a restaurant, demand has the following behavior: in the morning, the demand is 10 customers an hour, in the afternoon the demand is 15 customers an hour and, finally, in the nighttime the demand is 9 customers an hour. The average time that each table takes is around 90 minutes. The restaurant has 12 tables.

Calculate the performance measures and find the respective conclusions.

5. Let us assume that in a system the demand is 7 customers an hour. The service time per customer is 10 minutes and there are 2 servers. Get the number of servers required for the following situations:

a. We want there to be only enough servers in the system to serve the demand.
b. Let us assume that we want the probability of a customer having to wait not to be higher than 0.25.

Table 11. Increment of the demand

Base Demand	+10%	+20%	+30%	+40%
7 Customers an Hour	¿?	¿?	¿?	¿?

c. We want the system to always work at ¾ of its capacity.
d. Let us assume that the demand increases following the behavior shown in Table 11.

Repeat subsections a, b and c for each demand case (Table 11). How does the number of servers required for each policy change in respect of the increase in demand?

REFERENCES

Borst, S. C., & Seri, P. (2000). *Robust algorithms for sharing agents with multiple skills*. Working paper, CWI, Amsterdam.

Curry, G. L., & Feldman, R. M. (2011). *Manufacturing systems modeling and analysis* (2nd ed.). Berlin: Springer. doi:10.1007/978-3-642-16618-1

Gans, N., Koole, G., & Mandelbaum, A. (2003). Telephone Call Centers: Tutorial, Review, and Research Prospects. *Manufacturing & Service Operations Management: M & SOM*, *5*(2), 79–141. doi:10.1287/msom.5.2.79.16071

Gross, D., Shortle, J. F., & Thompson, J. M. (2008). *Fundamentals of queueing theory*. Hoboken, NJ: J. Wiley & Sons. doi:10.1002/9781118625651

Hillier, F. S., & Lieberman, G. J. (2005). *Introduction to operations research* (8th ed.). New York, NY: McGraw-Hill.

Kleinrock, L. (1975). *Queueing systems*. New York, NY: Wiley.

Taha, H. A. (2011). *Operations research: An introduction* (9th ed.). Upper Saddle River, NJ: Prentice Hall.

Taylor, H. M., & Karlin, S. (1984). *An introduction to stochastic modeling* (3rd ed.). Orlando, FL: Academic Press.

Chapter 4
Single-Stage Non-Markovian Systems and Variability

ABSTRACT

The variability of a queueing system is the main theme of Chapter 4; some of the approaches that have been developed to obtain the performance measures are presented. Part of the material is devoted to variability due to equipment failures and their effect on performance measures, as well as in estimating the required capacity. The chapter ends with the topic on propagation of variability in a queue and the respective approximations to calculate the variability at the exit.

INTRODUCTION

The previous chapter presented a set of relationships that enable us to predict the behavior of a work station by using demand, service time and the number of parallel work stations as information.

Likewise, Little's Law is a general equation that relates the amount of work in system with the service demand and residence time in the system. However the administrators often observe that there is an increase in the amount of work in process and the cycle time in the processes and systems under their responsibility, even when there has been no change in the system's average service time or the average demand. How many administrators manage processes where the performance gradually deteriorates as the months go by?

DOI: 10.4018/978-1-5225-5264-2.ch004

As asserted by Hopp and Spearman (2009), one reason for this inefficiency is variability in the system. This chapter will present a set of models for approximating the performance measures of a queue that fails to comply with the condition of having service times and/or times between arrivals of an exponential nature where it is necessary to have information about the dispersion of the system's service time.

POLLACHECK-KINTCHINE APPROXIMATION FOR THE CYCLE TIME

We must remember that, in the M/M/1 case, the evolution of the system is defined by the number of customers present in instant t and, in consequence, the stochastic process $\{N(t), t > 0\}$ is a Markov chain. For M/G/1 systems, the $\{N(t), t > 0\}$ process is not a Markov chain as, when there is a customer in the system and a new arrival increases the number of customers in the system, $N(t) \geq 1$, the service time that has elapsed (residual time) for the customer that was already there has to be considered, which does not conform with the lack-of-memory property of Markov processes. So, the state diagram does not apply in this case (Stewart, 2009).

Let us assume that demand λ follows a Poisson distribution and that the service discipline is First In, First Out and independent of the service time; let us also assume that for the service time, the only specification is for it to follow a general probability distribution (G) with mean t_S and standard deviation σ_S and for the station to have a single server. Under Kendall's notation, these types of systems are M/G/1 and M/G/c. The time spent in queue for an M/G/1 system is obtained using the following expression:

$$CT_q = \left(\frac{1+C_S^2}{2}\right)\left(\frac{\lambda t_S}{1-\lambda t_S}\right)t_S \qquad (1)$$

where the parameter $C^2_S = \text{var}_S / (t_s)^2$ is the squared coefficient of variation or simply the variability. We must bear in mind that C^2_S is a measure of dispersion, while (1) is known as a Pollaczek-Kintchine Formula and, just like the Markov systems, the system has to be stable, in other words, $\rho = \lambda t_S < 1$.

The average work in process of the system is:

$$WIP_S = \lambda t_S + \frac{\left(\lambda t_S\right)^2 + \lambda^2 \sigma^2{}_S}{2\left(1 - \lambda t_S\right)} \tag{2}$$

Or alternatively:

$$WIP_S = \left(\frac{1 + C_S^2}{2}\right)\left(\frac{\rho^2}{1 - \rho}\right) + \rho \tag{3}$$

Observe that if $C^2{}_S = 1$, then formulas (1) and (3) are the relationships for Markovian systems.

Example 4.1: Let us assume that a cafeteria has a single person providing the service: they take the order, prepare the coffee and charge. The average service time is 7 minutes per order or, in capacity terms, they attend to 9 customers an hour, this time is exponentially distributed, so the standard deviation is 7 minutes. The manager makes some adaptations to the operation performed by the cashier and, in a new sampling, the standard deviation for the service time is 3.5 minutes. Get the performance measures and compare the performance of the system before and after the adaptations that made the standard deviation for the service time decrease.

The customers arrive at a rate of 8 customers an hour and the time between every customer is exponentially distributed.

Observe that when the service time is exponentially distributed (M/M/1 case), then variability is $C^2{}_S = (7)^2 / (7)^2 = 1$, and the cycle time is 1 hour. Whereas in the M/G/1 case, variability is $C^2{}_S = (3.5)^2 / (7)^2$ which shows less dispersion; while the equation P-K predicts a cycle time of 0.678 hours (40.6 minutes), as we can see in Table 4.1. Likewise, the average work in process in the M/M/1 case is 7.33, whereas, in the M/G/1 case, the work in process is 5.424.

In general, as the variance of the service time decreases, the value of $C^2{}_S$ is reduced, which shall be reflected in the system by a reduction of the cycle time in the system and work in process? On the other hand, if $C^2{}_S$ increases, in other words, if the dispersion of the service time rises, then the cycle time and work in process will also increase.

Table 1. Comparison of an M/M/1 system with an M/G/1 system

Scenario	Before Adaptations	After Adaptations
Performance measures	M/M/1	M/G/1
Congestion	0.88	0.88
CT_s(hr.)	1	0.678 (40.6 min)
WIP_s	7.33	5.424

G/G/1 and G/G/c Queues

Unlike Markov systems where the cycle time equations are exact, there are only equations to approximate the cycle time for G/G/1 and G/G/c systems. The main difficulty in getting exact equations lies in the fact that the lack-of-memory condition is not met, whereas the literature proposes several equations, known as approximations. Said expressions require the calculation of service variability (C^2_S), variability of arrivals (C^2_a), variance of arrivals (σ^2_a) and service variance (σ^2_S).

Kingman's Approximation of Cycle Time for G/G/1 Systems

During the operation of a queue, it is common to observe periods when demand increases, thus congesting the system. In these situations of high congestion where ρ is very close to 1, the time in queue has been proven to follow an exponential distribution. John Frank Charles Kingman developed the following formula to approximate the time in queue of a G/G/1 system where ρ is very close to 1:

$$CT_q \approx \left(\frac{C_a^2 + C_S^2}{2}\right)\left(\frac{\rho}{1-\rho}\right)t_S \tag{4}$$

Expression (4) is known as Kingman's formula and some of this formula's characteristics are (Whitt, 1993):

1. If the time between consecutive customers is exponentially distributed, then $C^2_a = 1$and the Polazcek-Kintchine formula is obtained.
2. It is exact for the M/M/1 system.

3. If $C^2_a = 1$and the service time is exponentially distributed, then (4.4) is considered to be an approximation for the G/M/1 system.
4. Approximation is better for systems with high congestion: $\rho \to 1$.
5. The term $(C^2_a + C^2_S)/2$ is a weighting of the arrival and service variabilities.

Approximations for G/G/c Systems

Based on the fact that the time in queue follows an exponential distribution in situations where ρ is very close to 1, Kingman's formula gives us the following improvement for systems with several parallel servers:

$$CT_q \approx \left(\frac{C_a^2 + C_S^2}{2}\right)\frac{\rho}{c(1-\rho)} \tag{5}$$

Approximation (5) is not recommended as it gives a considerable deviation for very high values of c. Authors such as Whitt (1993) and Allen (1990) propose a general approximation for the time in queue using the term $(C^2_a + C^2_S)/2$,thus taking into account the variability of arrivals and service time:

$$CT_q \approx \left(\frac{C_a^2 + C_S^2}{2}\right)\left[CT_{M/M/c}\right] \tag{6}$$

The term $\left[CT_{M/M/c}\right]$ is the approximation of the time spent in the queue for Markov systems (not the exact formulas seen in chapter 3). H. Sakasegawa, 1977, proposes the following formula for M/M/c systems (Whitt, 1993):

$$CT_q \approx \frac{\rho^{\sqrt{2(c+1)}-1}}{(1-\rho)c} t_S \tag{7}$$

Allen (1990) gives the following approximation based on (6) and (7) for the time in queue in G/G/c systems:

$$CT_q \approx \left(\frac{C_a^2 + C_S^2}{2}\right)\left(\frac{\rho^{\sqrt{2(c+1)}-1}}{(1-\rho)c}\right)t_S \tag{8}$$

Equation (8) is known as the Allen-Cunneen approximation. Its main characteristics are:

1. It is exact for M/M/c systems.
2. If $C^2_a = 1$, (8) is an approximation for M/G/c systems.
3. It is generally considered to be a good approximation.

There are other formulas, but they have the characteristic of incorporating new parameters and more calculations. A spreadsheet can help to solve this situation when you want to apply these formulas to decision-making. The papers of Allen (1990) and Whitt (1993) will serve as a reference for those readers that are interested in finding out more about the performance of some approximations.

Example 4.2: Let us consider a work station where the service time is 5.45 minutes with a standard deviation of 3 minutes. The customers arrive at the work station every 6 minutes with a standard deviation of 2 minutes. Get the average waiting time in the queue, the average cycle time in the system, the average work in process in the system and the size of the queue. The station has a single server.

By applying (7) we get the cycle time as 3.75 minutes in the queue. The rest of the properties are obtained by applying Little's Law and the results are given in Table 2.

Table 2. Performance measures

Congestion	0.908
CT_q	11.18 minutes
CT_S	16.63 minutes
WIP_q	1.86 customers
WIP_S	2.77 customers

Variability

As we have already mentioned, variability is a measure of dispersion that indicates the amplitude of the standard deviation in respect of the mean.

Figure 1 shows the normal probability plot with mean x and two different values of σ. When we compare both graphs we can see that the extreme values have a higher probability of occurrence in the graph on the left-hand side than in the graph on the right-hand side. The squared coefficient of variation $C^2$is higher in the graph on the left-hand side than in the graph on the right-hand side.

What is the relationship between variability and the queueing systems? Let us assume that both figures represent the service time in two stations that manufacture the same product. One intuitive idea is that a service time with low variability is preferable in order to ensure that the probability of a customer staying a long time in the system is also low.

Variability is a way of measuring the deviation from the periodic or stable operation of a process. Locating the point and identifying the causes of this deviation will enable an administrator to make better decisions about the systems they are in charge of. Variability affects a system, as is evident, but the point is not only to quantify it but also to quantify its effect on the performance of a system.

As can be observed in example 4.1, variability has a relevant role to play in the study of the phenomenon of queueing. Seeking to maintain quality of service, companies implement actions to make the residence time or cycle time of a customer (a piece, order of purchase, etc) as low as possible.

Figure 1. Two cases with the same mean x = 5 a). σ = 2, b).σ= 1

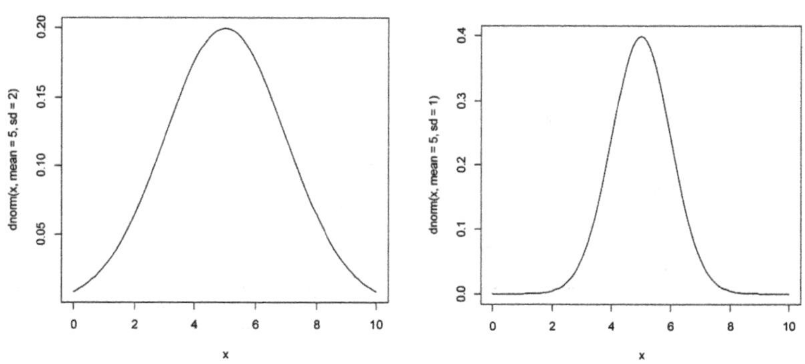

(a) (b)

Example 4.3: Let us now consider example 4.2. Applying (8) to several values of C^2_S, we calculate the time in queue and work in queue (Figure 2).

As can be appreciated, the variability of service time is in direct proportion to the time in queue. When C^2_S =0.33 the time in queue is 11.18 minutes and the length of the queue is 1.82 customers, whereas when C^2_S = 1, the time in queue increases to 30 minutes and the work in queue grows to 5 customers. Finally, when C^2_S = 2.5, the time in queue is 70 minutes and the work in queue jumps to 11.75 customers.

As can be appreciated in example 4.2, a system with high variability generates an accumulation of work in queue and, consequently, cycle time increases.

Every production and service system has elements that are subject to variability: demand, process time, repair time, time between failures, performance. A customer is averse to waiting, in consequence if they have two options to choose from, they will always prefer those systems where the queue of work in process is short because, from their perspective, their waiting time will be short. Reducing the variability of demand makes it possible to keep a small number of raw materials stored and lower the cycle time of a production order.

Classification of Variability

An administrator who is responsible for a system must understand what the causes of variability in the system are in order to be able to quantify it. According to Hopp and Spearman there are several sources of variability (Hopp and Spearman, 2009).

Figure 2. CT_q and WIP_q vs C^2_S

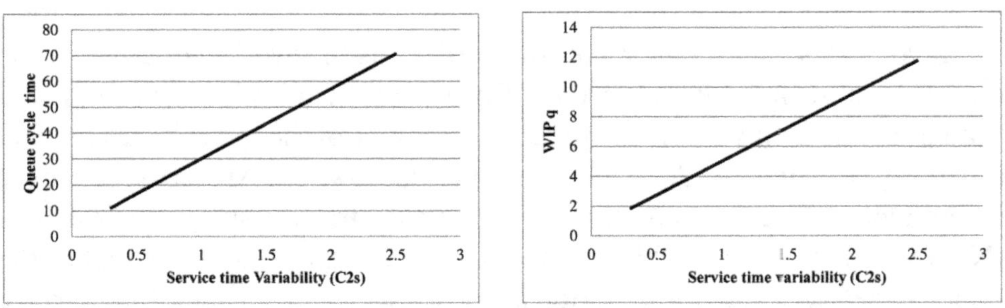

Natural Variability

This is the variability that is typical of the operation in a station. Take, for example, a cashier in a cafeteria or restaurant, who performs a series of similar, standard tasks when dealing with most of their customers: They ask which service is required, deliver the product or perform the job, charge, give change back, say thank you and attend to the next customer. In this case, the variability arises from the difference between customers (some will order more products, pay in cash, by card or make an extra payment) or, as the cashier tires, they will gradually be slower at performing these tasks.

In the approximations of the cycle time and work in process that we have already explained, we only took natural variability into account, using the parameter C^2_S.

Uncontrollable Variability

That which causes the production or service in the station to be interrupted. When an item of a machine fails, a customer's manufacturing or service process is interrupted, then the necessary time for finishing the process will be prolonged while waiting for the repair to be finished. Moreover, customers usually keep on entering the system, lengthening the queue and, in turn, increasing the waiting time by the length of time they need to wait in the queue (3).

In the examples given so far, the stations do not break down and are always available, this is an ideal situation because in real life machines have failures causing interruptions of production and consequently lowering the equipment's availability. The availability is a dimensionless number a = (0 – 1) is calculated using the following expression:

$$a = \frac{MTTF}{MTTF + MTTR} \tag{9}$$

where *MTTR* is the average repair time, *MTTF* is the average operating time until the catastrophic failure occurs. Equation (9) is also known as inherent availability as it only considers the characteristics of the system's design and reliability, therefore it does not consider shutdowns arising from changes in product, preventive maintenance or administrative issues or situations (NASA, 2015). The availability of the equipment can be used to calculate the effective service time of the station and the effective variance of the service

Figure 3. Effect of failures on cycle time and work in process

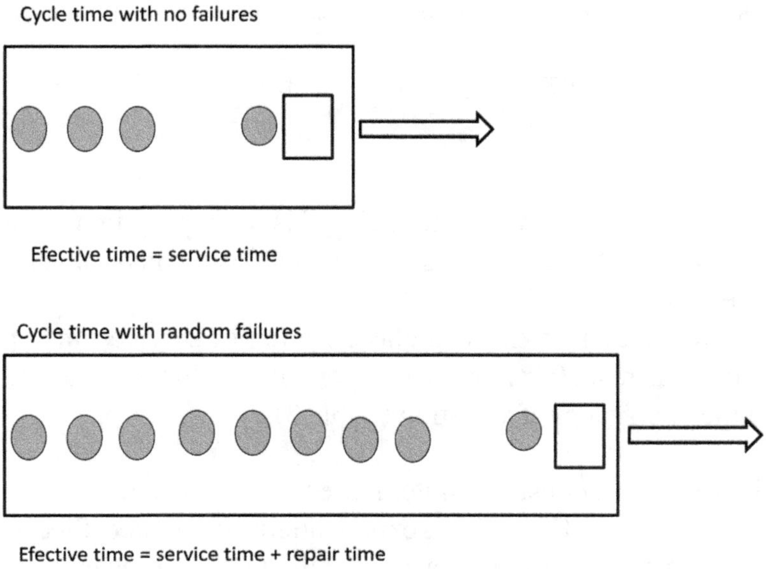

time (Hopp and Spearman, 2009, Curry and Feldman, 2009, Creemers and Lambrecht, 2011):

$$t_e = \frac{t_S}{a} \tag{10}$$

$$\sigma_e^2 = \left(\frac{\sigma_S^2}{a}\right) + \frac{\left(MTTR^2 + \sigma_{MTTR}^2\right)\left(1-a\right)}{aMTTR} t_S \tag{11}$$

where σ_e^2 is the effective variance; t_e is effective service time; t_S is the service time without failures; σ_S^2 the variance in the service time and σ_{MTTR}^2 the variance in the average repair time.

The resulting variability of the catastrophic shutdowns is the quotient obtained by dividing the effective variance by the square of the effective time:

$$C_e^2 = \frac{\sigma_e^2}{t_e^2} \tag{12}$$

By combining (10) and (11), Hopp and Spearman (2008) as well as Creemers and Lambrecht (2011) propose the following expression:

$$C_e^2 = C_S^2 + a\left(1-a\right)\frac{MTTR}{t_S} + C_{MTTR}^2\left(1-a\right)a\frac{MTTR}{t_S} \tag{13}$$

where C^2_e is the effective variability; C_{MTTR}^2 is the repair time variability; t_S is the average service time of the station and C_S^2 is the natural variability of the service time.

The following table presents a classification of service time variability (Hopp and Spearman 2008). It must be pointed out that this classification is based on the coefficient of variation (Table 3).

Example 4.4: Let us consider an item of equipment where the service time per piece is 6 minutes, which is exponentially distributed. Pieces arrive in the system at the rate of one every 20 minutes. The time is exponentially distributed. The equipment has failures that stop production and behave as shown in Table 4.

Table 3. Classification of variability

Classification	Coefficient of Variation	Typical Example or Situation
Low	$C_s < 0.75$	Inherent to the process or operation: Tiredness on the part of the operator, rest breaks.
Medium	$0.75 \leq C_s < 1.25$	Moderate: breaks for changes of product.
High	$1.25 \leq C_s$	Failures that stop production.

Table 4. Time taken for repairs (hours)

0.75	2	3.5	6
1	2	4	8
1	2.5	4	8
1	3	4	9
1.25	3	5	9
1.5	3	5.5	14
1.5	3	6	16
1.75	3.25	6	144

An analysis of the data for the repairs gives the results shown in Table 5.

As can be appreciated, the repairs variability is classified as high. The data table shows that on one occasion the failure took 144 hours to repair, as a consequence, there is a wide dispersion of repair times.

We have the data about the times between failures shown in Table 6.

Likewise, the basic statistical analysis of the time between failures is as shown in Table 7.

In this case we observe that the variability of the times between failures is moderate. If we observe the data in the table we can see that, on one

Table 5. Descriptive statistics of the failures

Time to Repair (MTTR)	
Mean	8.85
Standard deviation	24.92
Variance of the sample	621.39
Coefficient of variation	2.81

Table 6. Time between failures (hours)

1	4	8
1	4	8
1.5	4	8
2	4	8
2	4.5	10
2.25	5	18
2.75	5.75	20
3	6	
3	6	
3.5	6	

Table 7. Descriptive statistics of the time in operation until the failure

Time Between Failures (MTTF)	
Mean	5.60
Standard deviation.	4.54
Variance of the sample	20.69
Coefficient of variation	0.812

occasion, the equipment operated for up to 20 hours. Now we shall proceed to calculate the availability:

$$a = \frac{5.6}{5.6 + 8.85} = 0.387$$

Effective service time:

$$t_e = \frac{t_S}{a} = \frac{6\,\text{min}}{0.387} = 15.5\,\text{min}$$

Effective variability:

$$C_e^2 = C_S^2 + a(1-a)\frac{MTTR}{t_S} + C_{MTTR}^2(1-a)a\frac{MTTR}{t_S}$$

$$C_e^2 = (1) + 0.387(1-0.3887)\frac{8.85}{6} + (2.813)^2(1-0.387)(0.387)\frac{8.85}{6}$$
$$C_e^2 = 4.12$$

Although the service time is exponentially distributed, on taking the failures into account, the service time goes from 6 minutes (which would be the ideal or design time) to 15.5 minutes. Failures cause the real or effective variability to be 4.12. On looking at the behavior of the failures a second time, we can see that the equipment shutdown for longer than it is in production. The performance measures, taking shutdowns resulting from equipment failures into account, are given in Table 8.

Table 8. Comparison of results

Performance Measures	Excluding Failures	Including Failures
ρ	0.30	0.775
WIP_s	0.4285	7.56
WIP_q	0.1285	6.79
CT_s	8.57 minutes	151.3 minutes
CT_q	2.57 minutes	135.82 minutes

As can be appreciated, the effect of failures on a production system signifies a loss of capacity that could result in a severe problem.

Example 4.5: Let us assume that the management takes the necessary actions to avoid another 144-hour shutdown. The descriptive statistics are as shown in Table 9.

The availability is $a = 0.55$. The effective service time is 6 min/ 0.55 = 10.82 minutes, the effective variability is

$$+ 0.55(1 - 0.55)(4.5/6) + (0.822^2)(1 - 0.55)(0.55)(4.5/6) = 1.31. \quad (1)$$

The results for the system's behavior are shown in Table 10.

By eliminating a prolonged repair time, the equivalent of reducing variability, the immediate effect is an increase in equipment availability. There is an evident reduction in cycle time in this example, as it is from 151 minutes to 25.55 minutes.

As has been demonstrated, it is always a good idea to consider possible failures in any systems being analyzed, as otherwise we run the risk of underestimating the performance measures.

Table 9. Descriptive statistics

Repair Time	
Mean	4.5
Standard deviation	3.702
Variance of the sample	13.708
Coefficient of variation	0.822

Table 10. Performance measures

ρ	0.541
WIP_s	1.277
WIP_q	0.736
CT_s	25.55 minutes
CT_q	14.73 minutes

Example 4.6: Let us consider a system with a single server that works under the operating conditions shown in Table 11.

Get the graph for cost vs number of servers in the case that excludes failures, with MTTR= 8 exponentially distributed hours and MTTR=20 exponentially distributed hours.

The operating cost of a station, assuming that it does not break down, is \$9.76 and the optimum number of servers is 3. In the case with MTTR= 8 hours, the optimum number of servers is 3 and the cost of \$10.47. There is a difference of 7.2% between the operating costs excluding failures vs operating costs including failures. When MTTR= 20 hours, the optimum number of servers is 4 and the operating cost is \$11.47 (Figure 4).

If the catastrophic shutdowns are prolonged, the consequence is a loss in capacity, higher variability and increased operating costs for the system.

Table 11. Operating data

Time Between Pieces (Minutes)	11 exponentially distributed
Service Time (Minutes)	7.5 exponentially distributed
MTTF (Hours)	100 σ= 10 hours
Server Cost (\$/Customer)	2.5
Waiting Cost (\$/Customer)	12.5

Figure 4. Operating cost vs number of servers for three MTTR cases

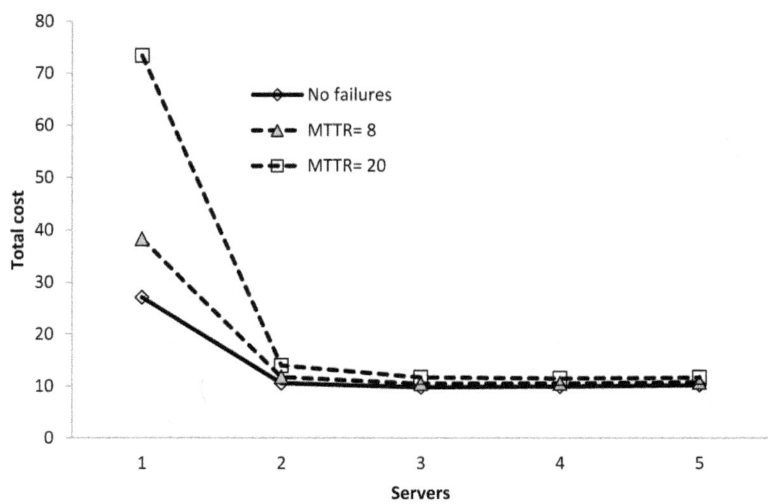

Controllable Variability

This arises from situations that depend on the decisions of the person in charge or the administrator of the system. The average time required to prepare the machine after a change of product; the estimated average time for preventive maintenance jobs or the average time that a doctor must be absent after serving a certain number of patients are some examples of these types of interruptions.

To get the effective time for these types of interruptions and their variability, we shall assume that:

1. The period or length of the interruption does not depend on the prior service history.
2. The period of interruption is a random variable with probability function $f(x)$, mean t_{setup} and variance σ^2_{setup}.

The effective time is:

$$t_e = t_S + \frac{t_{setup}}{O} \tag{14}$$

where O is the average number of production orders or the average number of pieces in a production run. The effective variance is:

$$\sigma_e^2 = \sigma_S^2 + \frac{\sigma^2_{setup}}{O} + \frac{O-1}{O^2} t_{setup} \tag{15}$$

And the effective variability is:

$$C_e^2 = \frac{\sigma_e^2}{t_e^2} \tag{16}$$

Example 4.7: Let us assume that the service time of an operation in a factory is 15 minutes and exponentially distributed. Product changes imply inspection, the cleaning and addition of attachments and take 12 minutes with a standard deviation of 7 minutes. The average size of a production run is 100 pieces. What is the variability of the system?

The effective service time (equation 14), variance (equation(15)) and variability (Equation (16)) are, respectively:

$$t_e = 15\,\text{min} + \frac{12\,\text{min}}{100pz} = 15.12$$

$$\sigma_e^2 = \left(15\right)^2 + \frac{7^2}{100} + \frac{100-1}{100^2}\left(12\right) = 237.37$$

$$C_e^2 = \frac{237.37}{15.12^2} = 1.03$$

Propagation of Variability

So far we have shown how variability is classified and its effect on the system's performance measures: when variability (either of the service time or of the times between consecutive arrivals of customers) increases, the cycle time and WIP also increase, signifying a loss of capacity in the system.

But what happens to the output of a station? The answer is that there is also variability but it has not been quantified. Output variability is not "immediately apparent" but can be quantified.

Let us consider a G/G/1 station. Customers arrive at the station separated by random spaces of time that follow a distribution with mean t_a and variance σ^2_a. Moreover the service time is a random variable with mean t_S and variance σ^2_S. The scheduling policy is First In, First Out.

The customers will leave the station every so often which is a random variable with mean t_d and variance σ^2_d. The output variability will be:

$$C_d^2 = \frac{\sigma_d^2}{t_d^2} \tag{17}$$

One way of obtaining output variability would be to take samples of the times between outputs, getting the mean and variance and then applying (17). As can be appreciated, both the variability at entry to the station and the service time variability contribute to the output variability . The degree of contribution of each factor shall depend on the degree of congestion of the station.

Marshall (1968) proposes the following expression for approximating the output variability of a G/G/1 station (Whitt, 1984):

$$C_d^2 \approx C_a^2 + 2\rho^2 C_S^2 - 2\rho(1-\rho)\frac{CT_q}{t_S} \quad (18)$$

where the term *CTq* in equation 18 is an approximation like Kingman approximation (4) or the Allen-Cunnen expression (8) explained above. For example, using the Kingman approximation for the cycle time in G/G/1 systems:

$$C_d^2 \approx \rho^2 C_S^2 + C_a^2(1-\rho) \quad (19)$$

In the case of the Allen-Cunnen approximation, we get the following expression for G/G/c systems:

$$C_d^2 \approx C_a^2 + 2\rho^2 C_S^2 - \rho\left(\frac{C_a^2 + C_S^2}{2}\right)\left(\frac{\rho^{\sqrt{2(c+1)}-1}}{c}\right) \quad (20)$$

Expression (20) is equivalent to the one proposed by Whitt (1984) and Hopp and Spearman (2008).

Example 4.8: From example 4.5 we know that the effective time is 10.82 minutes and the effective variability is 1.31. The demand is exponentially distributed and the station has a single server. By applying (19), we get:

$$C_d^2 \approx 0.545^2(1.31) + (1)(1-0.545) = 0.844$$

In this case, instead of using congestion and natural variability, we substitute in the effective congestion and the effective variability.

When a station works at a high level of congestion then the output variability tends to be similar to the service time variability of the station; on the other hand, if the station works at a low level of congestion then the output variability tends to be similar to the variability of the times between consecutive customers arriving at the station.

Example 4.9: Let us consider a work station where $C^2_S = 2$ and $C^2_a = 1$. Figure 6 shows C^2_d vs congestion using (19).

Example 4.10: From example 4.4, we know that service time variability, taking equipment failures into account, is 4.12. The graph shows the behavior of C^2_d. As the equipment's congestion increases, variability C^2_e predominates over C^2_a for the arrivals.

In other words, a work station with a high level of shutdowns and also a service demand that causes a high level of congestion are difficult to control. Given that the output of finished product or satisfied customers is irregular, an administrator will find it hard to achieve a good level of customer satisfaction.

Exercises

1. Let us consider a customer service center for the phone company. In the cashiers' area, there are 4 people working in parallel. The service time is 10 minutes with a standard deviation of 2 minutes. The time between consecutive customers is exponentially distributed and is 7.5 minutes with a standard deviation of 7 minutes on paydays; is 5.5 minutes with a standard deviation of 2 minutes on any other day of the month.

Let us assume that there are no differences between the cashiers. a. Get the performance measures.

Figure 5. Propagation of variability

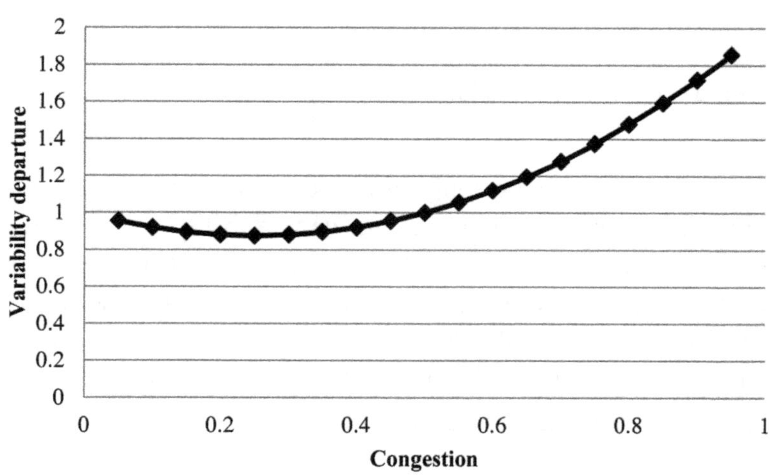

Figure 6. Propagation of variability for a case that includes failures in the equipment

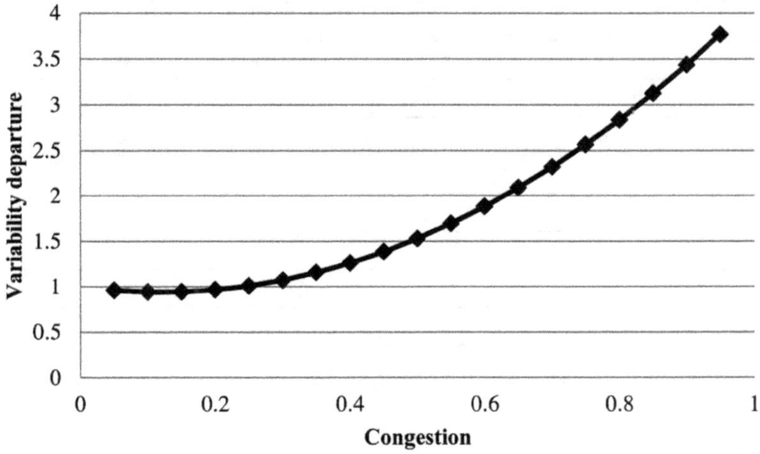

2. For the following data groups, get the coefficient of variation and classify them as per Tables 12-14.
3. In a factory, a work station has the operating data shown in Table 15.
    a. Get the performance measures for the case that excludes failures
    b. Get the performance measures for the case that includes failures with MTTR= 18 and σ_{MTTR} = 18.
    c. Get the service time that minimizes the cost, with the results of subparagraph b

Table 12. Group 1

23.88	2.92
16.50	1.71
4.04	11.92
0.44	9.34
5.14	4.33
3.99	3.69
40.4	3.72
2.54	31.34
15.06	7.35
11.03	36.17

Table 13. Group 2

37.54	41.23
41.49	40.90
39.76	38.94
40.60	40.74
39.88	40.69
40.02	39.63
38.36	40.81
40.23	39.85
38.66	40.57
41.86	37.68

Table 14. Group 3

1.59	24.39
12.41	12.29
6.27	38.53
13.61	3.12
4.44	2.59
2.37	3.16
13.02	6.53
35.97	13.12
10.83	7.09
3.25	27.97

Table 15. Operation data

Time Between Pieces (Minutes)	5 exponentially distributed
Service Time (Minutes)	2.0 exponentially distributed
MTTF (Hours)	50 σ= 10 hours
Server Cost ($/Customer)	7.5
Waiting Cost ($/Customer)	9.0

d. Let us assume that MTTR is dropped to 10 hours and is exponentially distributed. What is the service time that minimizes the operating cost?

4. Calculate the output variability for the systems shown in Table 16-19.

Table 16.

Time Between Pieces (Minutes)	3 exponentially distributed
Service Time (Minutes)	2.0 exponentially distributed

Table 17.

Arrival Rate (pcs /hr)	6, Poisson arrivals
Service Time (Minutes)	5, σ= 6 minutes

Table 18.

Arrival Rate (pcs /hr)	10, Poisson arrivals
Service Rate (pcs / hr)	11, σ= 5 pcs/hora

Table 19.

Time Between Pieces (Minutes)	5 exponentially distributed
Service Time (Minutes)	2.0 exponentially distributed
MTTF (Hours)	50 σ= 10 hours
MTTR (Hours)	18 σ= 18 hours

REFERENCES

Allen, A. O. (1990). *Probability, statistics, and queueing theory: With computer science applications* (2nd ed.). New York: Academic Press.

Buzacott, J. A., & Shanthikumar, J. G. (1993). *Stochastic models of manufacturing systems*. Upper Saddle River, NJ: Prentice Hall.

Creemers, S., & Lambrecht, M. (2011). The Modeling of Interrupts and Unplanned Absences in Health Care Operations. *Supply Chain Forum: An International Journal, 12*(1), 32-40.

Curry, G. L., & Feldman, R. M. (2011). *Manufacturing systems modeling and analysis* (2nd ed.). Berlin: Springer. doi:10.1007/978-3-642-16618-1

Hopp, W. J., & Spearman, M. L. (2008). *Factory physics*. Long Grove, IL: Waveland Press.

NASA. (Ed.). (n.d.). *Availability prediction and analysis*. Retrieved November 28, 2016, from http://oce.jpl.nasa.gov/practices/at3.pdf

Whitt, W. (1984). Departures from a Queue with Many Busy Servers. *Mathematics of Operations Research, 9*(4), 534–544. doi:10.1287/moor.9.4.534

Whitt, W. (1984). Approximations for departure processes and queues in series. *Naval Research Logistics Quarterly Naval Research Logistics, 31*(4), 499–521. doi:10.1002/nav.3800310402

Whitt, W. (1993). Approximations for the Gi/G/M queue. *Production and Operations Management, 2*(2), 114–161. doi:10.1111/j.1937-5956.1993.tb00094.x

Chapter 5 Open Queueing Networks

ABSTRACT

Systems often have two or more stages and a customer must go several stages before finalizing their service. The systems can be arranged in series or in network. Chapter 5 is dedicated specifically to the performance analysis of systems that have several stages both in series and those that have network arrangement; the theorems of Burke and Jackson are presented; the calculations of the flow and variability in a network and of the measures of performance (cycle time and work in process) are also exposed. Several codes are proposed in Scilab Language to perform calculations automatically. The chapter ends with a section devoted to identifying and analyzing the bottleneck in a system from a cost approach.

MULTI-STAGE MODELS

In the earlier chapters, we looked at queues under the assumption that the system consists of a single stage, in other words, when the customer leaves the station, its transformation process is finished and it does not need to go through a further stage, so it leaves the system.

However a customer usually requires several stages to complete its transformation process or sometimes has to return to one of the stations to repeat a process, in other words, to undergo a reprocess. There are a variety of systems where the customers may start its process through any of the stations, not necessarily through the one identified as station 1 and neither is it not necessary for the customers go through all the stages as they can also leave the system through any of the stations that make up the network (hence their name, open networks).

DOI: 10.4018/978-1-5225-5264-2.ch005

In the analysis of the performance and administration of a system formed by queues with several stages, it is vital to use indicators that enable us to establish its behavior, in order to give the decision-makers elements to support their decisions, such as increased capacity, the effect of reworks or the result of changes in demand. When making decisions, managers, administrators and, in general, everyone who is responsible for a system focuses on specific performance measures, for example:

1. **Total Work in Process (1):** The average number of customers in the system, which is the sum of the accumulated work of each station.

$$WIP_{Total} = \sum_{i=1}^{n} WIP_i \tag{1}$$

2. **Cycle Time (2):** The period from when the customer enters the first stage of the process until it leaves the system. The time that an order takes, from the moment it enters the system until the product is delivered, is a much-used measurement for service quality.

$$CT_{Total} = \frac{WIP_{Total}}{\sum_{i=1}^{n} \gamma_i} \tag{2}$$

3. **Production Rate or Throughput (3):** The number of pieces produced per unit of time. In the case of lines that do not have any constraint on the number of customers that can stay in the queue

$$Th = \sum_{i=1}^{n} \gamma_i \tag{3}$$

Systems in Series and M/M/C Stations

In systems such as factories, hospitals or restaurants, the customer's transformation process cannot be performed in a single stage as the piece or customer has to flow through the stations and be submitted to a transformation process in each station.

Let us consider a factory that has three stages, where a different operation is carried out in each stage. The pieces enter exclusively through station one; as soon as the processing time is finished they move on to machine 2; when the respective processing time ends, the pieces are sent to machine 3; then, as soon as they come out of this machine, they leave the system (Figure 1).

Now, let us consider Figure 2 where the process has two stations. The customers arrive at the system and their service starts in station 1. Once the service is completed, customers enter station 2 and when that process ends, the customers leave the system. The mean processing time in each machine is t_{s1} and t_{s2} respectively with a standard deviation of σ_{1s} and σ_{2s} respectively.

The mean time between outputs from station 1 is t_{d1} and has a standard deviation of σ_{1d}. Output variability is C_{d1}. The flow from station 1 becomes the demand flow for station 2, in other words: $C_{dl} = C_{a2}$.

To initiate the study of this class of systems, we will assume the following:

1. The customers only enter the system through the first station; the time between consecutive customers is a random variable that follows an exponential probability distribution.

Figure 1. Serial system

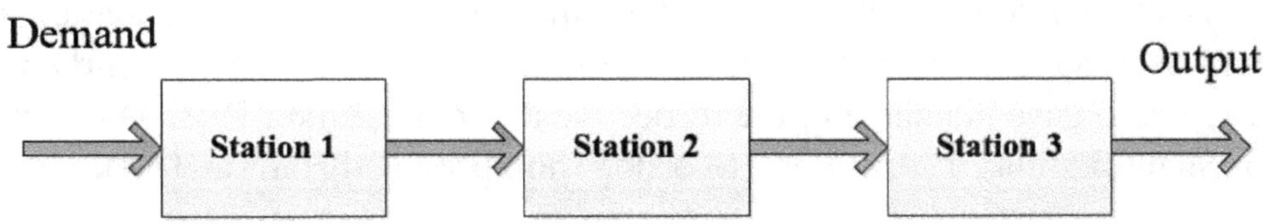

Figure 2. Variability propagation

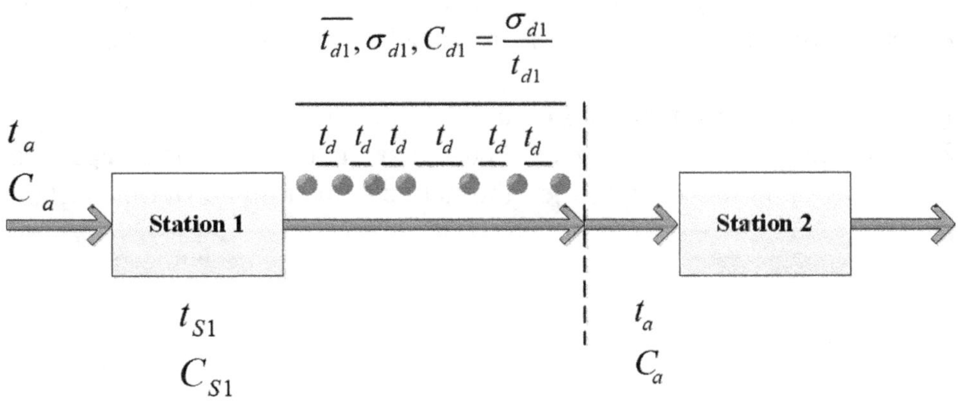

2. The service time in each station is a random variable with mean t_s and follows an exponential distribution.
3. The customer or piece leaves the system exclusively through the last station.
4. Reprocesses are not admitted. The customer or piece cannot return to any of the stages.
5. There is no constraint on the number of customers that can stay in the system.

When a customer leaves the station, one of the following cases happens: a. It leaves the station empty or b. It leaves the station with at least one customer occupying the system.

Let us assume that when a customer finishes its service, it leaves station 1 empty; the time t_{d1} that elapses until the next customer leaves will have a probability distribution that is the sum of two exponentially distributed random variables (assumptions 1 and 2), so the time between outputs will also be exponentially distributed.

Now let us assume that when a customer leaves station 1 and at least one other customer remains there, the time that elapses until the next output will have a mean of $1/\mu$ and will also be exponentially distributed. Assuming that the output from station 1 is sent to station 2, and if the service time in station 2 is exponentially distributed, then the time between outputs from station 2 will be exponentially distributed. The above was demonstrated by Burke in 1956 who gave his name to the respective theorem (Burke, 1956, Buzacott & Shanthikumar, 1993, Gross, Shortle, Thompson and Harris, 2008).

Burke's Theorem

Let us consider a system with M/M/1and M/M/c stations with an arrival rate λ. Let us assume that the system is in a steady state (Figure 3). Then:

1. The outputs from the stations are Poisson with a rate λ.
2. In each instant t, the number of customers in the system is independent of the order in which they left the station in the instants of time prior to t.

Figure 3. Burke's theorem

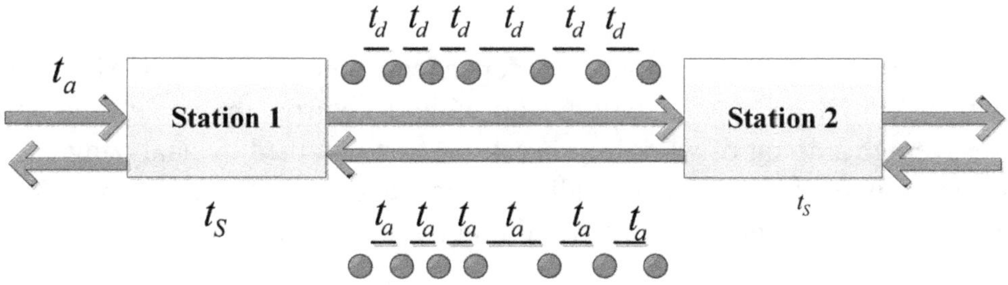

Corollary

The number of customers in a system cannot be determined solely on the basis of the departures from an M/M/1 station, in other words, although an observer may have seen a "lot" of customers leave the system in instant t_{i-1}, this does not make it possible to calculate or estimate how many customers there are at that time inside the system in an instant of time t_i (this is an event without any memory). Neither is it possible to have an estimate of the system's average service time that is solely based on the observation of the output process.

If the system consists of Markov stations, in series, with entry through station 1 and departure through station 2, then the system will behave exactly the same in the opposite direction; the departure of customers through station 2 will be the equivalent of customer entries, and the input flow through station 1 will be the equivalent of the departure of customers.

When the customers' inter-departure time follows an exponential probability distribution, then taking the flow in reverse, this will be the equivalent of to the time between arrivals and shall also follow an exponential distribution; given that the random phenomena that follow this distribution are memoryless, the state of the system (number of customers in the system) in instant t shall be independent of its state in the previous instants. Let (4) probability $P(N_i)$ of finding N customers in station i, then, for each station (Stewart, 2009):

$$\begin{cases} P\left(N_1 = k\right) = \left(1 - \rho_1\right)\rho_1^{\,k} \\ P\left(N_2 = l\right) = \left(1 - \rho_2\right)\rho_2^{\,l} \end{cases} \tag{4}$$

where $\rho = \lambda/\mu$. Then the probability of finding k customers in station 1 and l customers in station 2 will be:

$$P\left(N_1 = k, N_2 = l\right) = \left(1 - \rho_1\right)\left(1 - \rho_2\right)\rho_1^{\ k}\rho_2^{\ l} \tag{5}$$

According to (5) the stations act independently, therefore for a line in series with n M/M/1 or M/M/c stations and without returns, the cycle time and the average amount of work in process are first obtained by analyzing each station separately, then and overall cycle time and overall work in process of a system in series are obtained by adding the individual results.

Example 5.1: Let us consider a system consisting of two stages in series. At station 1 an average of 1 customer arrives every 10 minutes, the time between arrivals is exponentially distributed. In stage 1, there is a single machine and the average service time is 7 minutes whereas in stage 2, there is a single machine and the average service time is 8 minutes. In both cases the service time is exponentially distributed. Get the cycle time and the average work in process of each station.

By applying the equations for M/M/c systems, we get the following results (Table 1):

The total cycle time is the sum of the cycle time of each station (in series) 40 + 23.33 = 63.33 minutes. The WIP is calculated using Little's Law (6):

$$WIP = \frac{1}{10} \times 63.33 = 6.33 \tag{6}$$

Network Systems and M/M/C Stations (Jackson Networks)

The previous section studied systems in series: the stations or stages of the system are one after another, demand exclusively enters through the first station and satisfied customers leave the system through the last station.

Companies usually have equipment that performs a set of specialist operations and it is also common for said equipment to be grouped in well-

Table 1. Performance measures, example 1

Station	Congestion	Average Cycle Time (min)	WIP
1	0.7	23.33	2.33
2	0.8	40	4

defined sections. It is worth mentioning that the products are transported to each section by some system, depending on what product are being manufactured.

The order in which they visit each one of the stages of the process depends on each product, which may visit a machine or section more than once, not necessarily starting the transformation process at the station identified as station 1 (Figure 4).

We will also observe that there is an input flow from the outside towards the inside of the system, which can be through one or more stations; it will also be evident that the customers' path through the system depends on the service they require, so we will often see some of the customers leave the system through any one of the stations that make up the network.

In general terms, a system with the aforementioned characteristics is conceived as a network: the stations are connected to each other and every customer follows a path within the system. A system in series is a particular case of an open network.

For example, in manufacturing systems, these are called Workshop and among their advantages we can mention the fact that grouping the machines in a well-defined section gives the personnel more efficient control, thus increasing the efficient use of the machines and enabling the manufacture of several different types of products.

It is important to point out that these types of systems have disadvantages, such as having high levels of work in process or that requiring the assignment of priorities at each station to be continuously reviewed.

In order to estimate the average cycle time and the average work in process in a network, the number of customers in the system must first be estimated. Then, employing the decomposition method, the variability of output from each station is determined and this information used to calculate the performance measures of the network.

Figure 4. Network of queues

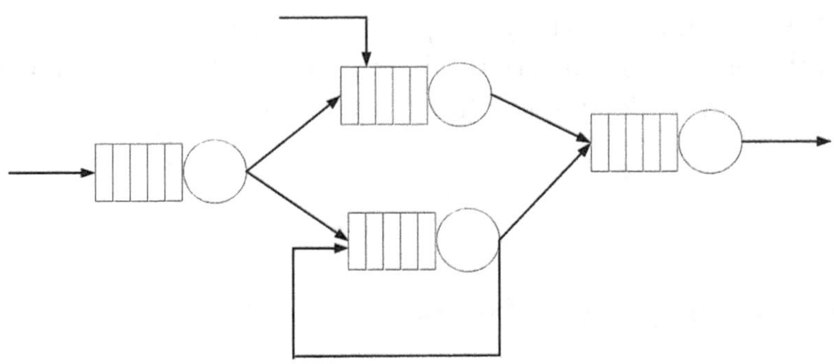

Flow in a Network

For the analysis of queueing networks, let us assume that:

1. The system has n work stations. Each station has 1 or more servers and there is no difference between them.
2. Customers enter the system through any one of the stations.
3. The service time in each station is t_s.
4. Once the service is finished, the customer leaves station i and goes on to the next stage j with probability p_{ij}, it can even leave the system through any one of the stations if process is completed.
5. There is no constraint on the number of customers that can stay in the queue.

The input and output flow balance equations show that the number of customers that circulate within a network is found by using the following equation:

$$\lambda_i = \gamma_i + \sum_{j=}^{n} \lambda_j p_{ji}; i = 1, 2, \ldots, n \tag{7}$$

Equation (7) establishes that the total input flow to a station i is the sum of the external flow plus proportion p_{ij} of the flow that leaves station j heading for station i. The solution of this system of linear equations is obtained using any known method for this case. For example, the above system has the following representation in matrix form:

$$\boldsymbol{\lambda} = \gamma + \boldsymbol{\lambda P} \tag{8}$$

where $\boldsymbol{\gamma}$ is the external flow vector and a known value; $\boldsymbol{\lambda}$ is the flow vector for inputs to the stations that make up the network that has to be calculated; and **P** is the probability matrix or process path matrix and is known. The system (8) has the following solution (Curry and Feldman, 2011):

$$\boldsymbol{\lambda} = \left(\boldsymbol{I} - \boldsymbol{P}\right)^{-1} \gamma \tag{9}$$

where **I** is an identity matrix of the same size as **P**.

Let us assume that the service time and the time between consecutive customers follow an exponential distribution. Let k_i be the number of customers in station i and the vector that defines the state of the system in a given instant be $(k_1, k_2, \ldots, k_n)$.

Jackson's Theorem (Jackson, 1957)

The probability P_k^i of finding k customers in station i is obtained by the following expression:

$$P_k^i = \begin{cases} P_0^i \dfrac{\left(\frac{\gamma_i}{\mu_i}\right)^k}{k!} & k = 0,1,\ldots,m_n \\ P_0^i \dfrac{\left(\frac{\gamma_i}{\mu_i}\right)^k}{m!}\left(m\right)^{k-m_n} & k = m_n, m_{n+1}, \ldots \end{cases} \tag{10}$$

Based on (10), the probability distribution of the states of the system described by the above expression has the following form:

$$P_k^i = \left(P_{k_1}^1\right)\left(P_{k_2}^2\right)\left(P_{k_3}^3\right)\ldots\left(P_{k_i}^i\right) \tag{11}$$

The model (11) establishes that if the system is in a stationary state, the stations that form the network will behave like systems that are independent of each other and the system of equations for the flow of customers in the network (also known as traffic equations) (9) satisfies the expression (11).

Example 5.2: Let us consider a system with 3 stations. The input flows to each station are as follows: 4 customers per hour through station 1, 2 customers per hour to station 2 and 2.25 customers at station 3. When the customer finishes its service at station 1, 20% go to station 2 and 50% go to station 3, the rest leave the system. All the customers that enter station 2 go to station 3 when they have finished their service. When the customer finishes their service at station 3, they leave the system (Figure 5).

Matrix **P** will show us the path the customers follow inside the system (12):

Figure 5. Network of queues with 3 stations

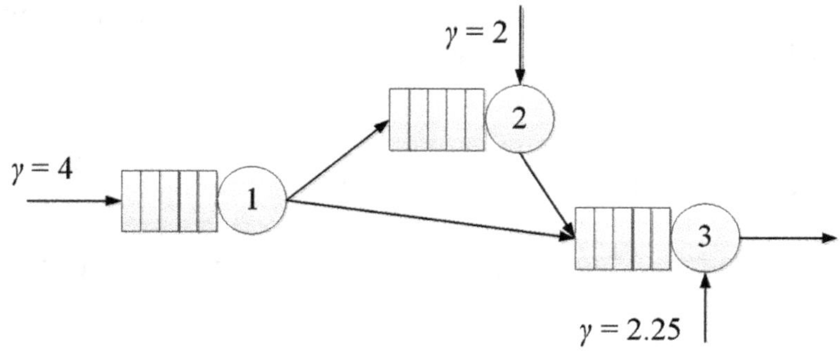

$$P = \begin{pmatrix} 0 & 0.2 & 0.5 \\ 0 & 0 & 1 \\ 0 & 0 & 0 \end{pmatrix} \tag{12}$$

The external flow vector is $\boldsymbol{\gamma}^t$ =[4, 2.8, 7.3].

The number of customers circulating through the system is: 5.97 customers enter station 1; 4.25 enter station 2 and 2.25 enter station 3.

Example 5.3: Let us consider a factory with 4 stations. The input flows to each station are: 3 pieces a minute through station 1 and 1 piece a minute at station 2. 10% of the pieces produced by machine 1 require a reprocess; 45% of the pieces are sent to station 2 and 45% to station 3; from station 2, all the pieces are sent to station 4; all the production of station 3 is sent to station 4. Of the material that is obtained in station 4, 5% is returned for reworking in the same machine and 1% is return to station 3. The process path is as follows (13):

$$P = \begin{pmatrix} 0.1 & 0.45 & 0.45 & 0 \\ 0 & 0 & 0 & 1 \\ 0 & 0 & 0 & 1 \\ 0 & 0 & 0.01 & 0.05 \end{pmatrix} \tag{13}$$

The average number of customers circulating through the system is: of 3.33 customers enter station 1; 2.5 customers enter station 2; 1.54 customers enter station 3 and 4.26 customers enter station 4.

From Burke's Law, we learn that the properties of queueing networks without reflows or returns to any of the stations can be obtained by analyzing every station separately, considering that the customers that leave each stage feed into the next station until they leave the system.

Corollary

For this, the time that elapses between consecutive customer arrivals needs to follow an exponential probability distribution. In other words, the arrivals must form a memoryless process, so the state of the system (the number of customers) will only depend on the service time.

When, for some reason, there is a return (for example a reprocess) and the customer has to visit a station more than once, the input flow into the station that receives said return cannot be treated as a Poisson process (Burke's Theorem does not consider returns).

On the other hand, Jackson's theorem establishes that for networks with M/M/1 or M/M/c stations, the entry of customers into each station follows a Poisson distribution and is not invalidated even in the case of a reflow or return to a station (Figure 6).

This leads us to conclude that, in order to get the properties of an open network even including reflows, the average number of customers that enter each station must first be determined; when the input flow has been determined then every station is analyzed in isolation to determine the average cycle time and the average amount of work in process.

Figure 6. Network of queues with markovian stations

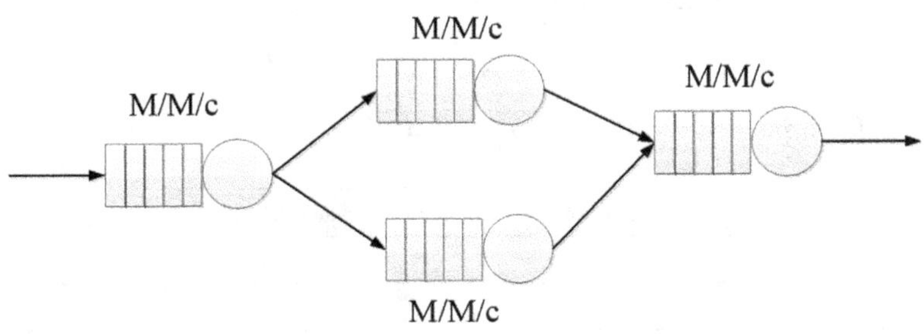

Example 5.4: In the case of the system in example 3, let us assume that the service times (in hours) of each station that forms the network are: 1/6.5 in station 1, 1/5.5 in station 2 and 1/3 in station 3. Let us assume that the service times follow an exponential distribution and that one customer at a time is served. The number of customers entering each station is: 5.97 customers enter station 1; 4.25 enter station 2 and 2.25 enter station 3.

The congestion at each station is: $\rho_1 = 0.918$, $\rho_2 = 0.857$ and $\rho_3 = 0.75$.

By applying the equations for M/M/1 systems, we get the following results in (14) (15) and (16):

$$CT_1 = \frac{1}{\mu - \lambda} = \frac{1}{6.5 - 5.97} = 1.88 \text{ hours} \tag{14}$$

$$CT_2 = \frac{1}{5.5 - 4.95} = 1.818 \text{ hours} \tag{15}$$

$$CT_3 = \frac{1}{3 - 2.25} = 1.33 \text{ hours} \tag{16}$$

The summary of the system's performance measures is given in the Table 2.

The total work in process in the network is: 11.22+8.99+2.99 = 23.2. 4 + 2.8 + 7.3 = 14.1 customers enter the system and, based on the flow balance, we can verify that 14.1 customers leave. The cycle time in the network is obtained by applying Little's Law once again and dividing the average amount of work in process by the amount produced by the system, in other words: 23.2 hours / (4+2.8+7.3)= 1.64 hours

Table 2. Performance measures, example 5.4

Station	Congestion	Average Cycle Time (hr)	WIP
1	0.918	1.88	11.22
2	0.857	1.81	8.99
3	0.75	1.33	2.99

Example 5.5: Let us consider the following queueing network with the following process path (17):

$$P = \begin{pmatrix} 0.1 & 0.5 & 0.3 & 0 \\ 0 & 0.2 & 0.2 & 0.6 \\ 0 & 0 & 0 & 1 \\ 0 & 0 & 0.4 & 0 \end{pmatrix} \tag{17}$$

An average of 4 pieces per hour enters the system through station 1. The consecutive inter-arrival times follow an exponential probability distribution. There is only one machine in each station and the average service times also follow an exponential probability distribution. The service times are shown in Table 3.

In this case the Networks.xls template will be used to get the properties of the network. The number of customers circulating in the network in each station is shown in Table 4.

The congestion of the system, cycle time and the work in process in each station is given in Table 5.

The total amount of work in process in the network is 19.324 pieces and the average cycle time in the network is 4.83 hours.

Table 3. Service time, example 5

Station	Average Service Time (hr)
1	0.182
2	0.286
3	0.200
4	0.143

Table 4. Flow in the network

Station	Flow (pz/hr)
1	4.444
2	2.778
3	4.259
4	5.926

Table 5. Performance of the network, example 5

Flow (pzs / hr)	Congestion	Average Work in Process	Average Cycle Time (hr)
4.444	0.808	4.211	0.947
2.778	0.794	3.846	1.385
4.259	0.852	5.750	1.350
5.926	0.847	5.517	0.931

Systems With G/G/C Stations

The above section consisted of a general review of queueing networks with M/M/1 and M/M/c stations. As soon as the number of customers that enter each station was obtained, the cycle time and the amount of work in process were calculated.

In the case where the service time and time between arrivals follow general distributions, we need to resort to approximations in order to get the properties. As we explained earlier, the flow that leaves a station becomes the demand flow of the next station, so the output variability needs to be calculated as this will be the input variability in the next stage of the process.

Chapter 4 explained a set of equations that approximate the value of a station's output variability and will be taken up again in the following sections.

Systems in Series

As in the Markov case, the presentation will start with studying systems in series. Let us assume that we have a system consisting of two stations in series, reintroducing the assumptions on page 7. Let us assume that the service time follows a probability distribution with a mean of $1/\mu$ and a standard deviation of σ. Likewise let us assume that the time between arrivals follows a probability distribution with a mean of $1/\lambda$ and a standard deviation of σ_a.

The cycle time and squared coefficient of variation of the output from a station are calculated using the expressions (18) and (19).

$$CT_S \approx \left(\frac{C_a^2 + C_S^2}{2}\right)\left(\frac{\rho^{\sqrt{2(c+1)}-1}}{(1-\rho)c}\right)t_S + t_S \tag{18}$$

$$C_d^2 \approx C_a^2 + 2\rho^2 C_S^2 - \rho\left(\frac{C_a^2 + C_S^2}{2}\right)\left(\frac{\rho^{\sqrt{2(c+1)}-1}}{c}\right) \tag{19}$$

By applying these expressions to station 1 of figure 2 we get expressions (20) and (21):

$$CT_{S,1} \approx \left(\frac{C_a^2 + C_{S,1}^2}{2}\right)\left(\frac{\rho_1^{\sqrt{2(c_1+1)}-1}}{(1-\rho_1)c_1}\right)t_{S,1} + t_{S,1} \tag{20}$$

$$C_{d,1}^2 \approx C_a^2 + 2\rho_1^{\,2} C_{S,1}^2 - \rho_1\left(\frac{C_a^2 + C_{S,1}^2}{2}\right)\left(\frac{\rho_1^{\sqrt{2(c_1+1)}-1}}{c_1}\right) \tag{21}$$

In this case, $C_a^{\,2}$ refers to the squared coefficient of variation of the flow from outside that enters through station 1. The flow of customers that leaves station 1 goes to station 2. The average time between arrivals will be $1/\lambda$ and the variability at output $C_{d,1}$ will now be the variability with which the customers enter station 2:

$$CT_{S,2} \approx \left(\frac{C_{d,1}^2 + C_{S,2}^2}{2}\right)\left(\frac{\rho_2^{\sqrt{2(c_2+1)}-1}}{(1-\rho_2)c_2}\right)t_{S,2} + t_{S,2} \tag{22}$$

$$C_{d,2}^2 \approx C_{d,1}^2 + 2\rho_2^{\,2} C_{S,2}^2 - \rho_2\left(\frac{C_{d,1}^2 + C_{S,2}^2}{2}\right)\left(\frac{\rho_2^{\sqrt{2(c_2+1)}-1}}{c_2}\right) \tag{23}$$

In (22) and (23) the value of $C_a^{\,2}$ is swapped for C_{d1}. Given that all the customers that enter station 1 continue their path towards the station 2, then $\rho_2 = \lambda t_{s2}$. After station 2 the squared coefficient of variation of the output from previous station is used. For example, for a 3rd station we would get (24) and (25):

$$CT_{S,3} \approx \left(\frac{C_{d,2}^2 + C_{S,3}^2}{2}\right)\left(\frac{\rho_3^{\sqrt{2(c_3+1)}-1}}{(1-\rho_3)c_3}\right)t_{S,3} + t_{S,3} \tag{24}$$

$$C_{d,3}^2 \approx C_{d,2}^2 + 2\rho_3^{\,2} C_{S,3}^2 - \rho_2\left(\frac{C_{d,2}^2 + C_{S,3}^2}{2}\right)\left(\frac{\rho_3^{\sqrt{2(c_3+1)}-1}}{c_3}\right) \tag{25}$$

After station 2 the calculation process is repeated for each one of the n stations that make up the line and can be generalized in the following recursive expressions (26) and (27):

$$CT_{S,i} \approx \left(\frac{C_{d,i-1}^2 + C_{S,i}^2}{2}\right)\left(\frac{\rho_i^{\sqrt{2(c_i+1)}-1}}{(1-\rho_i)c_i}\right)t_{S,i} + t_{S,i} \tag{26}$$

$$C_{d,i}^2 \approx C_{d,i-1}^2 + 2\rho_i^{\,2} C_{S,i}^2 - \rho_i\left(\frac{C_{d,i-1}^2 + C_{S,i}^2}{2}\right)\left(\frac{\rho_i^{\sqrt{2(c_i+1)}-1}}{c_i}\right) \tag{27}$$

where $C^2_{d,i-1}$ is the squared coefficient of variation of the flow from outside that goes to station i. Once the properties of station 1 are obtained, the information of C^2_d is used to calculate the cycle time of the next station. The analytical process described above corresponds to the decomposition paradigm: the original system is divided into subsystems, which are smaller and easier to handle, so as to analyze them separately

Example 5.6: Let us consider a system with three stations. The operating data are given in Table 6.

Table 6. Operation data of the line

Station	Average Service Time	Std. Dev.	C^2_s
1	7 min.	2.5 min.	0.128
2	6 min.	6 min.	1
3	8 min.	8 min.	1

Let us assume that a piece arrives at the line with a mean of 9 minutes and a standard deviation of 6 minutes and let us also assume that there is only one machine in each station. Get the properties of the system.

The average cycle time (28), the average amount of work in process (29) and variability of the output from station 1 (30) are:

$$CT_{S,1} \approx \left(\frac{0.444+0.128}{2}\right)\times\left(\frac{0.778^{\sqrt{2(1+1)}-1}}{\left(1-0.778\right)\times 1}\right)\times 7+7=14.01 \tag{28}$$

$$C_{d,1}^{2} \approx 0.444+2\times\left(0.778\right)^{2}\times 0.128-0.778\times\left(\frac{0.444+0.1128}{2}\right)\left(\frac{0.778^{\sqrt{2(1+1)}-1}}{1}\right)=0.253 \tag{29}$$

$$WIP_{1}=\frac{1}{9}\times 14.01=1.556 \tag{30}$$

In order to calculate the cycle time in station 2, the output flow from station 1 must now be taken as the demand. Expressions (31), (32) and (33) show the results:

$$CT_{S,2} \approx \left(\frac{0.253+1}{2}\right)\times\left(\frac{0.667^{\sqrt{2(1+1)}-1}}{\left(1-0.667\right)\times 1}\right)\times 6+6=13.52 \tag{31}$$

$$C_{d,2}^{2} \approx 0.444+2\times\left(0.778\right)^{2}\times 1-0.778\times\left(\frac{0.444+1}{2}\right)\left(\frac{0.778^{\sqrt{2(1+1)}-1}}{1}\right)=0.585 \tag{32}$$

In station 2, the average amount of work in process will be:

$$WIP_{2}=\frac{1}{9}\times 13.52=1.502. \tag{33}$$

Finally, the results for station 3 are given in (34), (35) and (36):

$$CT_{S,3} \approx \left(\frac{0.585+1}{2}\right) \times \left(\frac{0.889^{\sqrt{2(1+1)}-1}}{\left(1-0.889\right) \times 1}\right) \times 8 + 8 = 58.72 \tag{34}$$

$$C_{d,3}^2 \approx 0.585 + 2 \times \left(0.889\right)^2 \times 1 - 0.889 \times \left(\frac{0.585+1}{2}\right)\left(\frac{0.889^{\sqrt{2(1+1)}-1}}{1}\right) = 0.91 \tag{35}$$

$$WIP_3 = \frac{1}{9} \times 58.72 = 6.52 \tag{36}$$

Observe that the variability of the flow that arrives at station 1 would be classified as low and the variability is moderate at the output from the line. In other words the output flow from station 3 is more irregular than the original flow and therefore harder to predict.

Example 5.7: Rodríguez, González, Hernández and Hernández (2016) report the following data for the service time in a hospital (Table 7). Let us assume that all the patients who arrive at the desk are served by the "Triage" and then by the specialist doctors; also let us assume that in the "Triage" only their vital signs are taken and no level of care priority is assigned to the patient. The patients arrive at the desk every 5.56 minutes with a standard deviation of 4.83 minutes.

From the analysis, we observed that 3 doctors are not enough to meet the service demand. The results of considering 4 doctors in the third stage are given in Table 8.

Table 7. Healthcare system

Station	Servers	Average Service Time	Std. Dev.	C^2
Reception	1	1.323 min.	1.037 min.	0.614
"Triage"	1	4.177 min.	2.369 min.	0.32
Doctor	3	20.91 min.	19.36 min.	0.85

Table 8. Performance of a healthcare system

Station	Congestion	Average Cycle Time (hr)	WIP	Variability of Departures
Reception	0.238	1.6	0.289	0.746
"Triage"	0.75	10.84	1.94	0.506
Doctor	0.94	4.06	46	

The desk has a low workload, so the probability of finding the system empty is high; the combination of both factors results in the variability in the output from this stage being very similar to the variability of the input flow of the patients.

In the following "Triage" stage, the classification operation is carried out using a standard method. So the natural variability is low and the workload is higher in this station, resulting in the output variability being closer to the service time variability. Finally, there is high congestion (in fact this is where the bottleneck is) in the station corresponding to the doctors and, as a result, the variability of output is closer to the service time variability.

Network Systems

To get the performance measures, the number of customers that circulate between the stations must first be obtained by applying the flow equations; It is worth mentioning that said equations do not need to fulfill the assumption of exponential distribution either for the times between arrivals or for the service time but the system needs to be in a steady state, in other words $\rho < \lambda / c\mu$. Moreover, we need the information about variability in order to get the cycle time and the average amount of work in process (Gross, Shortle, & Thompson, 2008)

For the analysis of queueing networks with non-Markovian stations, we will now assume that the time between arrivals follows a probability distribution with a mean of t_a and a standard deviation of σ_a. Likewise the service time for each station follows a probability distribution with a mean of t_s and a standard deviation of σ_s.

Now that we have the mean and standard deviation, we can get the squared coefficient of variation. This parameter is necessary for determining the cycle time by applying the approximations described in chapter 4.

For the analysis of the flow through the network's stations we assume that said process consists of three elements (Gross, Shortle, & Thompson, 2008, Whitt, 1983):

1. **Superposition Process:** The number of customers that enter a station consists of one or more inputs that join to form a single flow (Figure 7). The variability on entry to a station will be approximated by assuming that:

    a. All the flows of customers that are sent to a station *i* are independent of each other and every flow is a renewal process
    b. The flows join to form a single input to the station. Given that the total flow that enters station *i* is a combination of renewal processes that are independent of each other, then, to get its value, we shall assume that said input flow into the station is also a renewal process.

Let us consider a node in a network. Said node is entered by $N_{ji}(t)$ customers in instant *t*. Let us assume that each arrival $N_{ji}(t)$ is a renewal process and that the arrivals are independent of each other. Let us assume that the total number of customers that arrive at a station is given by $N_i(t) = N_{0i}(t) + \ldots + N_{ki}(t)$ and let us assume that $N_i(t)$ is also a renewal process. The limit value (37) of the arrivals at a station *i* in instant t from station *j* is λ_{ji}; likewise the limit value for the variance (38) in the arrivals at station *i* from station *j* is var_{ji}:

Figure 7. Superposition process

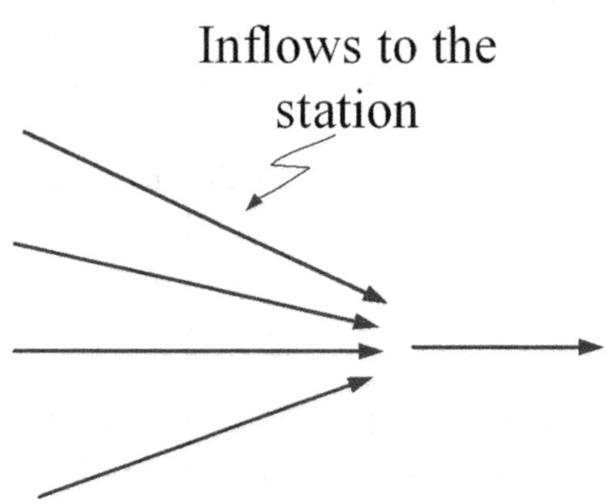

$$\lambda_{ji} \equiv \lim_{t \to \infty} \frac{\mathrm{E}\left[N_{ji}\left(t\right)\right]}{t} \tag{37}$$

$$\mathrm{var}_{ji} \equiv \lim_{t \to \infty} \frac{\mathrm{var}\left[N_{ji}\left(t\right)\right]}{t} \tag{38}$$

Given that the flows are independent processes, then:

$$\lambda_i = \lambda_{0i} + \ldots + \lambda_{ki} \tag{39}$$

$$\mathrm{var}_i = \mathrm{var}_{0i} + \cdots + \mathrm{var}_{ki} \tag{40}$$

Based on expressions (39) and (40), it can be demonstrated that the input variability of a station is approximated by using equation (41):

$$C_{a,i}^2 \approx \frac{\mathrm{var}_{ji}}{\lambda_i} \approx \frac{1}{\lambda_i} \sum_{j=0}^{n} \mathrm{var}_{ji} \approx \frac{1}{\lambda_i} \sum_{j=0}^{n} \lambda_{ji} {C^2}_{ji} \approx \frac{\gamma_i}{\lambda_i} C_{0,i}^2 + \frac{1}{\lambda_i} \sum_{j=0}^{n} \lambda_j p_{ji} {C^2}_{ji} \tag{41}$$

2. **The Queueing Process (Figure 8):** Where all the customers that enter a station are served. The variability in input flow and variability in service time define the station's output variability. For example, approximation (42) is used for a station with a single server or, for a system with several servers, we use (43):

Figure 8. Queueing process

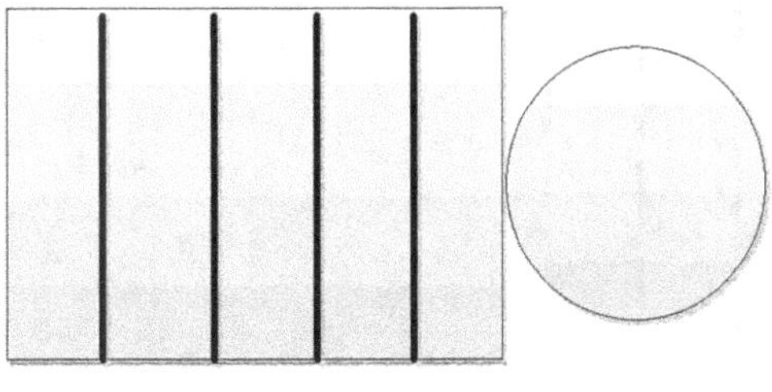

$$C_{d,i}^2 \approx \rho_i^2 C_{S,i}^2 + \left(1 - \rho_i^2\right) C_{a,i}^2 \tag{42}$$

$$C_{d,i}^2 \approx \left(1 - \rho_i^2\right) C_{a,i}^2 + \rho_i^2 \frac{C_{S,i}^2 + \sqrt{c_i} - 1}{\sqrt{c_i}} \tag{43}$$

3. **Division of the Flow (Splitting Process) (Figure 9):** When the customers leave the station, they follow their path as defined by the matrix of process path P, in other words, the output flow divides into one or more parts and each part goes to the next station, This includes reflows or reprocesses. The variability of each flow is approximated by using the following expression (44):

$$C_{ij}^2 = p_{ij} C_{d,i}^2 + \left(1 - p_{ij}\right) \tag{44}$$

Flow Variability

The above expressions can be combined to get a system of equations that enables us to calculate variability of the input to each station. First the approximation of the variability of the individual flows is substituted:

$$C_{a,i}^2 \approx \frac{\gamma_i}{\lambda_i} C_{0,i}^2 + \frac{1}{\lambda_i} \sum_{j=0}^{n} \lambda_j p_{ji} \left(p_{ji} C_{d,i}^2 + \left(1 + p_{ji}\right)\right) \tag{45}$$

Figure 9. Splitting process

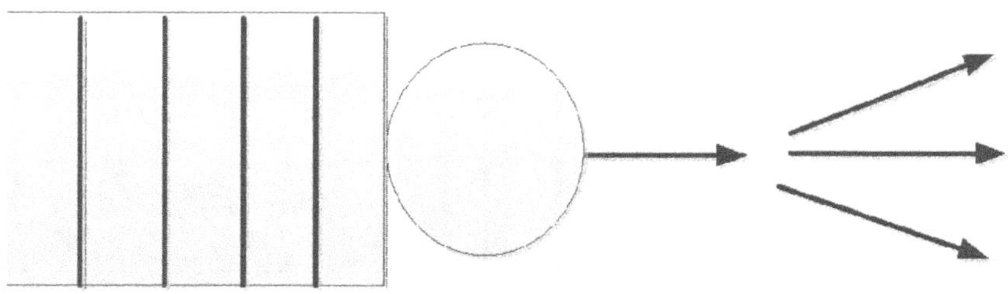

Then the approximation of the station's output variability is substituted in (45). In this case, the expression corresponding to a system with c servers will be used:

$$C_{a,i}^{2} \approx \frac{\gamma_i}{\lambda_i} C_{0,i}^{2} + \frac{1}{\lambda_i} \sum_{j=0}^{n} \lambda_j p_{ji} \left[p_{ji} \left[\left(1-\rho_j^2\right) C_{a,j}^{2} + \rho_j^2 \frac{C_{S,j}^{2} + \sqrt{c_j} - 1}{\sqrt{c_j}} \right] + \left(1 + p_{ji}\right) \right] \tag{46}$$

Expression (46) is applied to stations i =1,2,.., n. The first term is the variability resulting from the entry from outside; the second term corresponds to the contribution of each flow j=1,2,…,n that is superimposed to form a single flow that enters station i. To get obtain the input variability at each station we need to know the values C^2_{0i}, ρ, p_{ij}, c y λ and the system must also fulfill $\rho < \lambda/c\mu$. We have already explained the method for calculating the input flows at each station in section 5.3.1 . In this case it is also necessary to solve a set of simultaneous equations where the quadratic coefficients of variation for each station are the unknown factor.

Curry and Feldman (2009) propose the following approach as a system of simultaneous equations:

$$\boldsymbol{C}_a^2 = \left(\boldsymbol{I} - \boldsymbol{Q}^T\right)^{-1} b \tag{47}$$

where **I** is an identity matrix of n x n size; furthermore every entry of **Q**-Matrix is calculated as follows:

$$q_{ij} = \frac{\lambda_{ij} p_{ij}^2 \left(1 - \rho_j^2\right)}{\lambda_j} \tag{48}$$

Subscript j indicates the station whose flow is sent to station i. Each element of the vector **b** is calculated using the following expression:

$$b_i = \frac{\gamma_i}{\lambda_i} C_{0,i}^{2} + \sum_{j=1}^{n} \frac{\lambda_j p_{ji}}{\lambda_i} \left(p_{ji} \rho_j^2 \frac{C_{S,j}^{2} + \sqrt{c_j} - 1}{\sqrt{c_j}} + \left(1 - p_{ji}\right) \right) \tag{49}$$

Example 5.8: Let us assume that we have the following information about the process consisting of three stations. There are two input flows: one piece enters every 2 minutes through station 1, one piece enters every 0.75 minutes through station 2. Service time, and the SCV are shown in Table 9.

Stations 1 and 3 have a single server, station 2 has two servers. The path followed by the customers is given in the Figure 10.

The input flows to each station are 0.5, 1.6 and 1.33 pieces a minute. The calculations are given below for getting the squared coefficient of variation, by solving the system (47).

First, we get the values q_{ji} of **Q**-matrix by using the expression (48):

$$\begin{aligned} q_{12} &= \frac{0.5 \times 0.55^2 \left(1 - 0.75^2\right)}{1.605} = 0.0412 \\ q_{13} &= \frac{0.5 \times 0.45^2 \left(1 - 0.75^2\right)}{1.83} = 0.0242 \\ q_{23} &= \frac{1.605 \times 1^2 \left(1 - 0.8^2\right)}{1.83} = 0.3122 \end{aligned} \tag{50}$$

Table 9. Service time and variability of stations

Station	Average Service Time	C^2s
1	1.5 min.	1
2	1 min.	2
3	0.5 min.	0.75

Figure 10. Network of queues, example 5.8

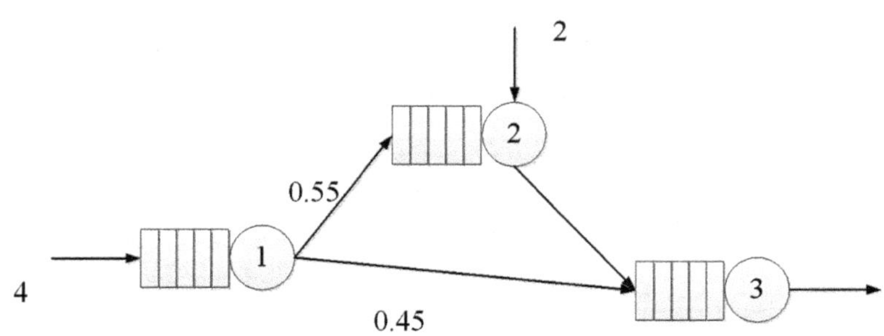

With the above results (50), the **Q**-matrix will be:

$$\boldsymbol{Q} = \begin{pmatrix} 0 & 0.0412 & 0.0242 \\ 0 & 0 & 0.3122 \\ 0 & 0 & 0 \end{pmatrix} \tag{51}$$

Now we replace the parameters in order to calculate the b vector given by the expression (49) thus getting the following results for b_1:

$$b_1 = \frac{0.5}{0.5}(1) + 0 = 1, \tag{52}$$

The inputs corresponding to b_2 and b_3 generate the following values:

$$b_2 = \frac{1.33}{1.605}(1) + \frac{0.5 \times 0.55}{1.605}\left(0.55 \times 0.75^2 \times \frac{1+\sqrt{1}-1}{\sqrt{1}} + 1 - 0.55\right) = 0.82 + 0.13 = 0.9588 \tag{53}$$

$$\begin{aligned} b_3 &= 0 + \frac{0.5 \times 0.45}{1.83}\left(0.45 \times 0.75^2 \times \frac{1+\sqrt{1}-1}{\sqrt{1}} + 1 - 0.45\right) \\ &+ \frac{1.605 \times 1}{1.83}\left(01 \times 0.80^2 \times \frac{1.991+\sqrt{2}-1}{\sqrt{2}} + 1 - 0.55\right) \\ b_3 &= 0 + 0.0988 + 0.96 = 1.059 \end{aligned} \tag{54}$$

Then the **b** vector, obtained using (51), (52), (53) and (54), ends up as follows:

$$\boldsymbol{C}_a^2 = \begin{pmatrix} 1 & 0 & 0 \\ 0.412 & 1 & 0 \\ 0.037 & 0.3122 & 1 \end{pmatrix}\begin{pmatrix} 1 \\ 0.958 \\ 1.059 \end{pmatrix} = \begin{pmatrix} 1 \\ 1 \\ 1.3958 \end{pmatrix} \tag{55}$$

Following on with the calculations, the $(\mathbf{I} - \mathbf{Q}^t)^{-1}$ matrix ends up as follows:

$$\left(\boldsymbol{I}-\boldsymbol{Q}^{t}\right)^{-1}=\begin{pmatrix}1 & 0 & 0\\ 0.412 & 1 & 0\\ 0.037 & 0.3122 & 1\end{pmatrix} \tag{56}$$

In the last step, the inverse of the matrix (56) is multiplied by the **b** vector (55) to get the quadratic coefficients of variation for the lambda flows:

$$\boldsymbol{C}_{a}^{2}\approx\begin{pmatrix}1 & 0 & 0\\ 0.412 & 1 & 0\\ 0.037 & 0.3122 & 1\end{pmatrix}\begin{pmatrix}1\\ 0.958\\ 1.059\end{pmatrix}\approx\begin{pmatrix}1\\ 1\\ 1.3958\end{pmatrix} \tag{57}$$

Once the quadratic coefficients of variation are obtained using (57) the approximations of G/G/c systems are applied to get the cycle time. For station 1, we get:

$$CT_{1}=\left(\frac{1+1}{2}\right)\left(\frac{0.75^{\sqrt{2\times1+2}-1}}{1\times\left(1-0.75\right)}\right)\times1.5+1.5=6\,\text{min}. \tag{58}$$

For stations 2 and 3:

$$CT_{2}=\left(\frac{1+1.99}{2}\right)\left(\frac{0.8^{\sqrt{2\times2+2}-1}}{2\times\left(1-0.8\right)}\right)\times1.0+1.0=4.25\,\text{min}. \tag{59}$$

$$CT_{3}=\left(\frac{1.3958+0.75}{2}\right)\left(\frac{0.915^{\sqrt{2\times1+2}-1}}{1\times\left(1-0.915\right)}\right)\times0.5+0.5=6.31\,\text{min}. \tag{60}$$

With cycle time obtained in (58), (59) and (60), the average amount of work in process is obtained by applying Little's Law to each station (Table 10).

The global cycle time is obtained by applying Little's Law again, $CT =$ 21.36 / 1.83 = 11.76 minutes.

The calculations can be speeded up by using a spreadsheet or building a program. The following code, written in Scilab, calculates the properties of open queueing networks with G/G/c stations (Scilab, 2012).

Table 10. Performance measures, example 8

Station	Congestion	Average Cycle Time (min)	WIP
1	0.75	6	3
2	0.8	3.79	6.09
3	0.915	6.31	11.76

```
function [x]=queue(n, gamm, c2g, P, ts, c2s, c)
    pt=P'
    I=eye(n,n)
    lamb=inv(I-pt)*gamm
        ro=lamb.*ts
        ro=ro./c
// Here starts the evaluation of SCV
    rocuad=(ro)^2
    unomenos=1-rocuad
    // qji elements
    for i=1:n
        for ii=1:n
            q(ii,i)=(lamb(ii)*(P(ii,i))^2)*unomenos(ii)/lamb(i)
        end
    end
    for i=1:n
        primert(i)=c2g(i)*gamm(i)/lamb(i)
    end
    for i=1:n
        for ii=1:n
qq(ii,i)=((lamb(ii)*(P(ii,i)))/lamb(i))*(P(ii,i)*rocuad(ii)*...
    (c2s(ii)+sqrt(c(ii))-1)/sqrt(c(ii))+1-P(ii,i))
        end
    end
  qq=qq'
  qq=sum(qq,2)
  b=primert+qq
  qt=q'
  qt=(I-qt)
  ca=inv(qt)*b
  //System performance measures
  ct=(ca+c2s)/2
  ct=ct.*(ro.^(sqrt(2*c+2)-1))
  ct=ct./(c.*(1-ro))
  ct=ct.*ts+ts
  //Litle's law
  wip=lamb.*ct
 [x]=[lamb,ro,ct,wip]
endfunction
```

As was mentioned at the start of the chapter, every product made in the workshop has a well-defined path. If said path does not have any reprocesses or else if the manufacture of a product implies or requires a single visit to each station, then the following analytical procedure will let us obtain the cycle time and the work for each one of the products manufactured in the workshop.

Example 5.9: Let us consider a plant where 2 products are manufactured. The data on the monthly demand for each material and the process path are given in the Table 11

Each station has the characteristics shown in Table 12.

The stations' service time follows an exponential distribution and the demand is also a Poisson process. As can be appreciated (Figure 11), the manufacturing process for products 1 and 2 only requires a single visit to each of the stages on its path.

The total material enters through station 1 and 90 pieces per hour are produced.

Following the path corresponding to product A, 33/80 go to station 2 from the output of station 1, then, 33/80 are sent from station 2 to station 3, finally 33/80 are sent from the output of station 3 to station 4. The **P**-matrix ends up as follows:

Table 11. Demand and process route

Item	Demand (pz/hr)	Route
A	33	1 – 2 – 3 – 4
B	47	1 – 3 – 5 – 2

Table 12. Service time and number of servers

Station	Average Service Time (hr)	Servers
1	0.025	4
2	0.035	3
3	0.03	3
4	0.04	3
5	0.02	2

Figure 11. Job shop for example 9

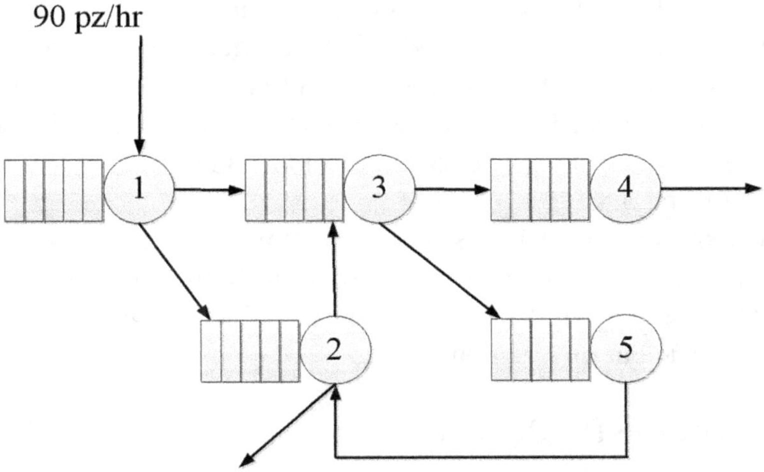

$$P = \begin{bmatrix} 0 & 33/80 & 47/80 & 0 & 0 \\ 0 & 0 & 33/80 & 0 & 0 \\ 0 & 0 & 0 & 33/80 & 47/80 \\ 0 & 0 & 0 & 0 & 0 \\ 0 & 1 & 0 & 0 & 0 \end{bmatrix} \tag{61}$$

The next step is to get the overall cycle time and WIP for the network, using (61). When the program is run, we get the results in Table 13.

The following table gives the analysis per product manufactured in the workshop. As can be appreciated in Table 13, at an overall level, station 2 is the one that has the most congestion in the system as the use of said station reached 93% of its capacity and we could consider it to be the bottleneck (also the cycle time is the highest of all the stations), followed by station 3

Table 13. Performance measures in the job-shop

Station	Congestion	Cycle Time (hr)	WIP
1	0.5	0.027	2.2
2	0.93	0.18	15.14
3	0.8	0.063	5.05
4	0.44	0.045	1.49
5	0.47	0.026	1.23

and station 1. The total amount of the system's work in process is 25.13 units. The overall cycle time is 0.314 hours or 18.84 minutes.

It is worth mentioning that a production manager might also be interested in cycle time per product. We observe that product A has the longest cycle time. It must be pointed out that the products compete for each one of the system's resources system. The service discipline in this example is assumed to be first in first out.

In the above examples, it was assumed that the equipment does not break down, which is an assumption that is hard to fulfill in reality. In the event of any failures, then we will have to calculate the effective service time and the effective coefficient of variation.

Analysis of the Bottleneck

This chapter has shown how to get the cycle time and the amount of work in process in queues in systems with G/G/c stations. The exercises highlighted the presence of stations with the highest level of congestion. Customers pile up at these stations more than at others that form part of the system and spend the longest there out of the total time they stay in the network. Said stations were called bottlenecks.

Throughout these chapters we have shown that a queue or lineup is a phenomenon that is inherent to systems: given that the transformation processes are not immediate and it is not possible to serve all the customers at the same time, it is normal for a queue to be formed with waiting material, in other words, a work-in-process inventory.

Sipper and Bulfin (1998) give a definition of a bottleneck as being:

... any operation that restricts the output of a system.

According to this definition, if the administrators want to increase the productivity of a system then they need to focus on the bottleneck. Other definitions for bottlenecks can be found in Chase and Aquilano (1989) or Krajewisky and Ritzman (1990).

In general, there is no formal definition of a bottleneck. Wang, Chen, Zhang and Huang (2016) identify up to 16 variants but the convention is that it is a station that limits or restricts the flow in a system.

Usually the station that limits production is assumed to be static, in other words, always the same, however also it has been demonstrated that the bottleneck is not static and can be moved from a station during the operation of a system. Considering this, we must point out that, in order to identify the bottleneck in a system, we should consider the operation's time horizon: short, medium and long term (Lawrence and Buss, 1995, Hu Zhiang, 2010).

Instead of a formal definition, Betterton and Silver (2012) propose a set of typical characteristics or elements:

1. It affects the residence time (cycle time) in the system.
2. It has the highest congestion.
3. It has the highest amount of work in process.
4. It causes a blockage in the previous stations and a lack of material in the following stations.
5. The frequency of blockages or lack of material does not happen as often as in the rest of the stations.
6. The proportion of time in repair or producing is higher compared with the inactive time resulting from a blockage or a lack of material.

The above characteristics are mainly focused on systems in series, likewise, d, e and f are very common in systems where the length of the queue for the stations is constrained to a certain number of spaces, in other words, buffer systems.

Methods

There are several methods for identifying a bottleneck. In two examples in this chapter we have used congestion level, which has a long-term approach. Betterton and Silver (2012) document some other methods for detecting the bottleneck: Active period (Roser, Nakano and Tanaka 2001), Inactive period (Sengupta, Das and Van Til, 2008), average production rate (Kuo, Lim, Meerkov, 1996), Point of return (Li, Chang, Ni, 2009) and Higher quantity of work in process (Lawrence and Buss, 1994).

In Lawrence and Buss (1995) apply an approach based on the first order conditions for studying the system's sensitivity to changes in the stations' service time. The bottleneck is the station with the highest impact or effect on performance. The perspective of this method is long-term operation using cost and gain as indicators. We will demonstrate the approach used in this method as it is based on Jackson networks and is of practical use for decision makers.

Economic Bottleneck in an Open Queueing Network

Let us assume that we have a network with stations connected to each other. A flow of customers enters the system and visit several stations before finally leaving the system.

The path followed by the customers is defined by a **P**-matrix and every input p_{ij} indicates the probability of a customer going from station i to station j. The service time follows an exponential probability distribution as does the time between consecutive arrivals at the stations. Let us assume that the stations have a single server and the service discipline is FCFS.

The operating cost of a network has the following general expression (Hillier and Lieberman, 2005):

$$\text{Cost=E[ServiceCost]+ E[WaitingCost]} \tag{62}$$

In (62) the service cost for a system with n stations is:

$$\mathrm{E}\left[\text{ServiceCost}\right] = \sum_{i=1}^{n} C_{Service,i} \times c \times \mu_i \tag{63}$$

Assuming that the waiting cost is proportional to the waiting time:

$$\mathrm{E}\left[\text{WaitingCost}\right] = \sum_{i=1}^{n} C_{WIPt,i} \times WIP_{S,i} \tag{64}$$

If the system has a single server and all the stations are M/M/1, combining (63) and (64) we get:

$$\operatorname{Cos} t = \sum_{i=1}^{n} C_{Service,i}\mu_i + \sum_{i=1}^{n} C_{WIP,i}\left(\frac{\lambda_i}{\mu_i - \lambda_i}\right) \tag{65}$$

The first term in (65) is the cost of the service and the second term is the total waiting cost, also known as the station flow cost. By applying first order conditions, we get optimum service capacities for the system:

$$\frac{dCostFlow}{d\mu_i} = C_{Service,i} - \frac{C_{WIP,i}\lambda_i}{\left(\mu_i - \lambda_i\right)^2}$$

$$C_{Service,i} - \frac{C_{WIP,i}\lambda_i}{\left(\mu_i - \lambda_i\right)^2} = 0 \tag{66}$$

$$\mu^* = \lambda_i + \sqrt{\frac{C_{WIP,i}\lambda_i}{C_{Service}}}$$

Expression (66) summarizes the process followed in chapter 3 where single-stage systems are explained. It must be remembered that in the case of M/M/1 systems, station i is independent of the rest of the stations in the system. Finding μ_i* does not imply that all the stations should work at the same service rate, which is why even when operating at optimum service rates there will be a station that restricts the flow through a network and there will be an accumulation of work-in-process inventory. Given that there is a service rate that ensures a minimum cost operation, an administrator should focus their decisions on controlling the station that restricts the flow and, in general, controlling the network's work-in-process inventory.

Trying to achieve a "zero-inventory" state would be a futile effort as it would require a station to be working at an infinite rate, a physical impossibility

Using cost as an indicator, we can identify the station that restricts the cash flow within the system.

One situation that the administrators need to deal with is cost management. When a system (such as a plant or a supply chain) is in operation, the capacity is usually modified to improve the flow through the system, so the question is: in which stations should the production capacity be increased to improve the flow? What will be the effect of these changes? If what we want is to improve the flow through the network, we only take from the gain expression (65) the term corresponding to the cost of the flow through the stations:

$$Cost = \sum_{i=1}^{n} C_{WIP,i}\left(\frac{\lambda_i}{\mu_i - \lambda_i}\right) \tag{67}$$

The decision variable is μ_i and, assuming that there is no difference between the stations, expression (66) will be derived in respect of $\mu_{i:}$

$$\frac{dCostFlow_i}{d\mu_i} = -\frac{C_{WIP,i}\lambda_i}{\left(\mu_i - \lambda_i\right)^2} \tag{68}$$

Expression (67) gives the rate of change in the cash flow cost per unit of change in the service rate. The above gives us the following definition:

Economic Bottleneck: *This is the station for which a change in the rate of service will generate the largest reduction in the cash flow cost; an analysis of the marginal cost enables us to find the station we should pay attention to from amongst a set of stations that form the network.*

Example 5.10: Let us consider a plant made up by four work stations with a single server each. The input flow is 10 pcs/hr and they enter through station 1 (Figure 12). The customers arrive following an exponential probability distribution. Every station's service times follow an exponential distribution and are shown in Table 14.

Table 14. Service time of example 10

Station	1	2	3	4
t_s (hr)	1/15	1/25	1/25	1/9

Figure 12. Network of example 10

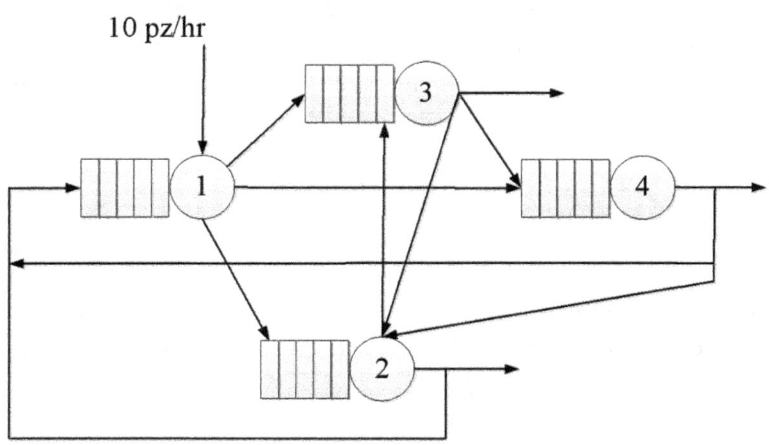

The path the customers follow is as follows (69):

$$P = \begin{pmatrix} 0 & 0.33 & 0.33 & 0.34 \\ 0.05 & 0 & 0.8 & 0 \\ 0 & 0.65 & 0 & 0.1 \\ 0.05 & 0.7 & 0 & 0 \end{pmatrix} \quad (69)$$

The input flow into each station is 11.37, 21.64, 21.1 and 5.9 customers per hour respectively. The congestion at each station is 0.75, 0.865, 0.84 and 0.65. In accordance with the usual definition of bottleneck, station 2 is the one that is restricting the flow through the network, therefore, according to this definition, our efforts to alter the service rate should center on this station.

Figure 13 shows the monetary impact of altering the each station's service rate by applying the equation for finding the derivative of the cost in respect of μ_i. We will get the highest reduction in the monetary cost if the service rate of station 3 is increased to 8%.

Figure 13. Cost rate of change

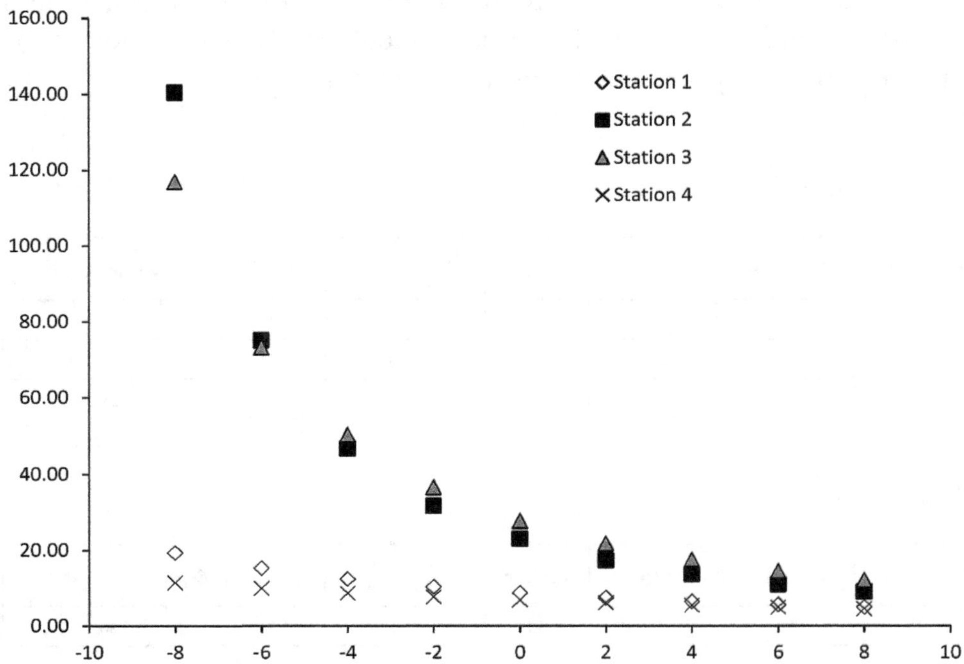

Exercises

1. A phone company has an office where a variety of administrative procedures are performed for the users. There is only one entrance. In stage R,the customers are channeled to Contracts (26%), Tills (40%) or Customer Care (13%). From the tills' area, 33.33% go to the Sales section or leave. From the Sales area, 20% go to the tills and the rest leave. From Customer Care, 30% then go to Specialist Care and 70% go to Sales. Finally, from the contracts area, 50% go to the tills and from Specialist Care, 33% go to the tills. The service times at each station are shown in Table 16.
    a. Let us assume that on a normal day, 25 customers arrive per hour. Get CT_S, CT_q, WIP_S and WIP_q in each service using data of Table 15.
    b. How many servers must be incorporated in each station so that in the contracts areas and in the sales area CT_S do not take longer than an hour?
    c. Let us assume that the costs of queueing at each station are shown in Table 16.

What is the economic bottleneck in this office? In other words, from a monetary point of view, in which station should the flow of customers be speeded up? Justify your answer using data of Table 16.

Table 15. Data

Station	Reception	Contracts	Cashier	Customer Attention	Sells	Especialized Attention
t_s (min)	1	25	5	15	15	20
C^2_s	1	1.5	0.75	1.25	1.75	1
Servers	1	3	2	2	2	2

Table 16. Waiting cost

Station	Reception	Contracts	Cashier	Customer Attention	Sells	Especialized Attention
C_{WIP}	10	50	50	75	100	120

2. Let us assume that 4 products are manufactured in a workshop. The process data for each case are set out in Table 17. The data for the service time are shown in Table 18. Let us assume that the service time and time between the arrivals of customers follow an exponential distribution.
   a. Get the workshop's performance measures. Which of the products has the longest cycle time?
3. In example 6 it is assumed that there are no equipment failures. Once again get the performance measures, assuming that we have the information shown in Table 19 about breakdowns at each station.

Table 17. Flow and rout of items

Item	Demand (pz/hr)	Route
A	25	1 – 2 – 3 – 4-5
B	25	2-3-4
C	15	5 – 3- 2 -1
D	35	3-4-1-5

Table 18. Service time and number of servers

Station	1	2	3	4	5
Service Time (hr)	0.025	0.035	0.03	0.04	0.02
Servers	5	3	3	3	2

Table 19. Failures

Parameter	1	2	3
a	0.9	0.8	0.9
MTTR(min)	15	15	15
C^2_{MTTR}	1	2	1.15

REFERENCES

Betterton, C. E., & Silver, S. J. (2012). Detecting Bottlenecks in Serial Production-Lines – a Focus on Interdeparture Variance. *International Journal of Production Research*, *50*(15), 4158–4174. doi:10.1080/00207543.2011.596847

Burke, P. J. (1957). The output of a queueing system. *Operations Research*, *4*(6), 699–704. doi:10.1287/opre.4.6.699

Buzacott, J. A., & Shanthikumar, J. G. (1993). *Stochastic models of manufacturing systems*. Upper Saddle River, NJ: Prentice Hall.

Chase, R., Aquilano, N. J., & Jacobs, R. (2000). *Operations Management for Competitive Advantage*. McGraw – Hill.

Curry, G. L., & Feldman, R. M. (2011). *Manufacturing systems modeling and analysis* (2nd ed.). Berlin: Springer. doi:10.1007/978-3-642-16618-1

Enterprises, S. (2012). Scilab: Free and Open Source software for numerical computation (OS, Version 5.3.3) [Software]. Available from: http://www.scilab.org

Gross, D., Shortle, J. F., & Thompson, J. M. (2008). *Fundamentals of queueing theory*. Hoboken, NJ: J. Wiley & Sons. doi:10.1002/9781118625651

Hillier, F. S., & Lieberman, G. J. (2005). *Introduction to operations research* (8th ed.). New York, NY: McGraw-Hill.

Hu, H., Jiang, Z., & Zhang, H. (2010). A Dynamic WIP Control Strategy for Bottlenecks in a Waffer Fabrication System. *International Journal of Production Research*, *48*(17), 5221–5233. doi:10.1080/00207540903117881

Jackson, J. R. (1957). Networks of Waiting Lines. *Operations Research*, *5*(4), 518–521. doi:10.1287/opre.5.4.518

Jackson, J. R. (1963). Job-Shop Like Queueing Systems. *Management Science*, *10*(1), 131–142. doi:10.1287/mnsc.10.1.131

Krajewsky, L. J., & Ritzman, L. P. (2001). *Operations Management: Strategy and Analysis*. Prentice – Hall.

Kuo, C. T., Lim, J. T., & Meerkov, S. M. (1996). Bottlenecks in serial production lines: A system-theoretic approach. *Mathematical Problems in Engineering*, *2*(3), 233–276. doi:10.1155/S1024123X96000348

Lawrence, S. R., & Buss, A. H. (1995). Economic Analysis of Production Bottlenecks. *Mathematical Problems in Engineering, 1*(4), 341–363. doi:10.1155/S1024123X95000202

Li, L., Chang, Q., & Ni, J. (2009). Data driven bottleneck detection of manufacturing systems. *International Journal of Production Research, 47*(18), 5019–5036. doi:10.1080/00207540701881860

Rodríguez, G. R., González, A., Hernández, S., & Hernández, M. D. (2017). Análisis del servicio del área de Urgencias aplicando teoría de líneas de espera. *Revista Contaduría y Administración., 62*(3), 719–732. doi:10.1016/j.cya.2017.04.001

Roser, C., Nakano, M., & Tanaka, M. (2001). A practical bottleneck detection method. *Proceedings of the 2001 winter simulation conference*, 949–953. doi:10.1109/WSC.2001.977398

Sengupta, S., Das, K., & Van Til, R. P. (2008). A new method for bottleneck detection. *Proceedings of the 2008 winter simulation conference*, 1741–1745. doi:10.1109/WSC.2008.4736261

Sipper, D., & Bulfin, R. L. (1998). *Production, Planning and Control.* Singapore: McGraw – Hill.

Stewart, W. J. (2009). *Probability, Markov Chains, Queues and Simulation.* Princeton University Press.

Taylor, H. M., & Karlin, S. (1984). *An introduction to stochastic modeling* (3rd ed.). Orlando, FL: Academic Press.

Wang, J., Chen, J., Zhang, Y., & Huan, G. (2016). Scheduled-based Execution Bottleneck Identification in a Job-Shop. *Computers & Industrial Engineering, 98*, 308–322. doi:10.1016/j.cie.2016.05.039

Whitt, W. (1983). The Queueing Network Analyzer. *The Bell System Technical Journal, 62*(9), 2779–2815. doi:10.1002/j.1538-7305.1983.tb03204.x

Chapter 6
Systems With Limited Capacity

ABSTRACT

In factories it is common that there are a limited number of spaces for the pieces; similarly, waiting rooms in a hospital can accommodate a limited number of people. Chapter 6 is dedicated to multi-stage systems where the stations have limited space for the queue; the chapter begins with the M/M/1/K systems and the calculation of its performance measures; the following is the bowl phenomenon and its implications in the efficiency of a system; then the M/G/1/K systems and the approaches developed to estimate the performance of this class of systems are presented. The chapter ends with the presentation of some optimization problems related to the M/M/c/K and M/G/c/K systems. Several codes are proposed in Scilab Language to perform calculations automatically.

SYSTEMS WITH LIMITED CAPACITY

In the models studied in the earlier chapters, we assumed that there is enough space for the customers to line up before a station; however in reality this assumption is not always the case. In a factory, usually the space for work-in-process inventory in front of a station is limited. In hospital waiting rooms or restaurants there is space for a certain number of customers. In a system such as stadium ticket offices or candy stalls in theaters and movie theaters, the area available for customers to line up is finite, however said area is so ample that it is generally not taken into account in the calculations.

DOI: 10.4018/978-1-5225-5264-2.ch006

Figure 1. A production line with buffers

Passageways and corridors in buildings or roads and highways are examples of systems that are represented as queueing systems with finite capacity. In these cases, the number of vehicles that can use the road (server) depends on the characteristics of the section (length), therefore the number of vehicles that can enter the system is restricted to a certain amount.

One indicator that is constantly watched by the administrators is the amount of work in process that accumulates before the stations, and usually maximum limits are set for the pieces lined up. The reasons for keeping or setting limits are a matter of space or economics.

An area where there is a fixed maximum number for waiting customers is normally easier to control than one where customers are allowed to arrive and line up. The work-in-process inventory must always be watched as it represents a monetary investment in raw materials.

When it is necessary to put a limit on the amount of work in process for any reason, this will trigger a series of effects that need to be considered, as they will have an impact on a system's indicators.

Let us consider a production line like the one in Figure 1. The system consists of n stations and there are m buffers. The customers enter through the first station and go through every stage until their process is finished in station n. Every machine consists of a single server and the process time in each station is $t_{s,i}$. There is K_i capacity in each station for the work in process: b_i places for customers in the queue plus the one being served.

What happens if all the available b places before a station i are occupied? The limitation on available space in a system will result in production shutdowns in the stations, which are then propagated backwards and forwards and whose effects increase the longer said shutdowns last.

For the station before machine i there is a phenomenon called blockage; this phenomenon is the fact that a customer cannot go on to the next stage in their process because there are is no place available in buffer j. This blockage is propagated to all the stations before machine i generating an accumulation

Figure 2. Buffer full and propagation of blocking

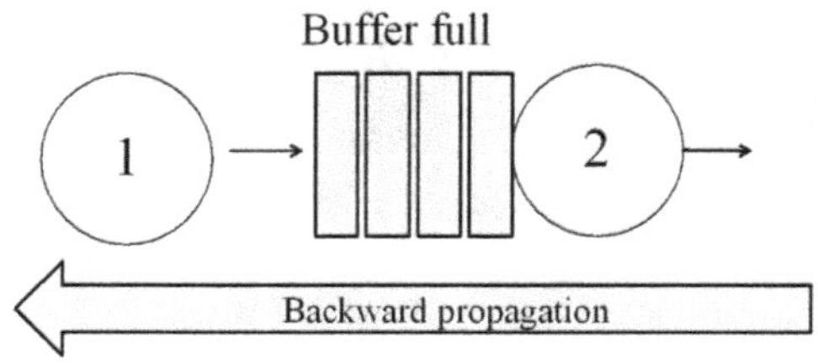

of material waiting to be processed that could cause the line to totally shut down (Figure 2).

Likewise, any lack of available material for the manufacturing process because of constraints of space combined with a slow machine or station results in a phenomenon called starvation, where the machines do not have any material to process and remain idle, and this is propagated forward (Figure 3).

There are two blocking mechanisms.

Blocking before service (Figure 4). When the buffer of station $i+1$ does not have any room for a new piece, then a signal is sent not to start processing the piece in station i.

Blocking after the process (Figure 5). In this case the piece is processed and, once it finishes the service, checks whether there is a place in the next

Figure 3. Propagation of blocking

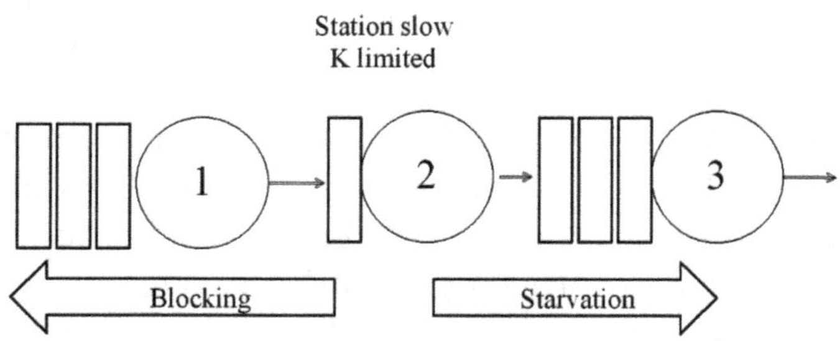

Figure 4. Blocking before service; The item cannot enter to the machine.

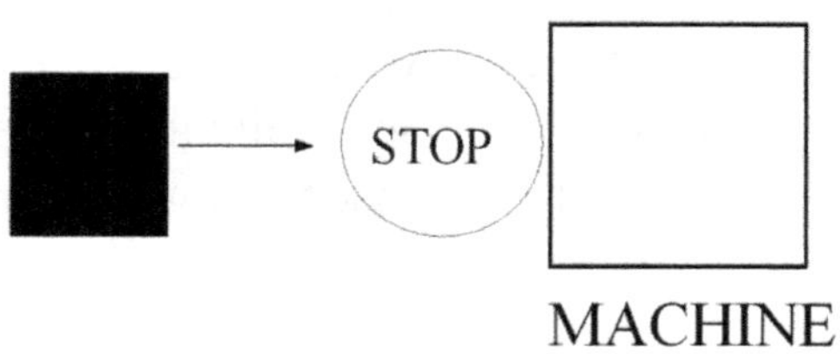

Figure 5. Blocking after service; Item cannot be released to the next station.

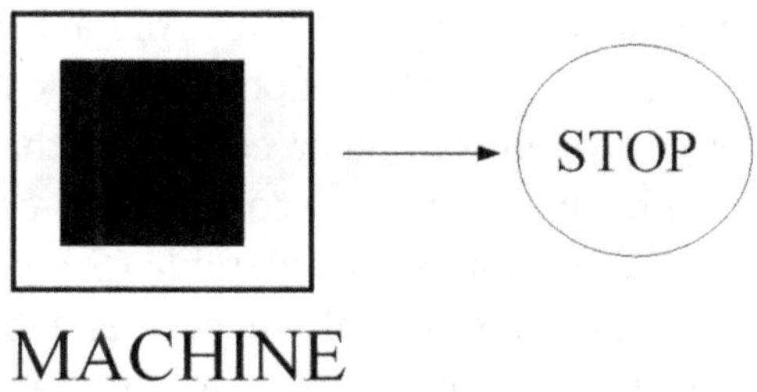

station. If there is one, then it leaves the machine, otherwise it stays on until at least one place is released.

Balsamo, Nitto-Persone and Onvural (2001) describe other variants for each of the types of blockage mentioned. In this book we will use the one corresponding to the BAS (blocking–after-service) to develop a model.

Calculation of the Properties of Systems in Series With M/M/1/K Stations

As with systems with unlimited capacity, there are methods for estimating the system's performance measures: cycle time, amount of work in process and output of the system. Let us assume that:

1. The stations' service time is $t_{s,i}$ and follows an exponential probability distribution
2. Demand exclusively arrives through first station and the inter-arrival times follow an exponential distribution.
3. The blocking–after-service mechanism is followed.

Exact Methods

As in the systems seen in chapter 3, the line can be modeled as a Markov chain. Let $X_i(t)$ be the number of customers that have been processed in station i-1 and whose transformation in station i was not completed in instant t. The set of states of this system is $\{X(t), t > 0\}$ and it is a continuous-time Markov chain where the set $X(t)= \{ X_1(t), X_2(t), \ldots, X_n(t) \}$ has all the possible states of the system as elements. Let the probability of finding the system in state x be: $p(x)= \lim t \to \infty$ P[$X_i(t)$ =x].

From the possible states of the system, we build the set of state equations where the decision variables are the probabilities of having n customers in the system (p_n); the point is that the number of possible states grows explosively, generating a very large system of linear equations that requires a lot of memory to store the information. In consequence the computational effort involved in obtaining the solution is considerable (Song, Takahashi, 1990). In view of all this, exact methods are only recommendable for systems with some stations. Some examples are reported in Altiok (1997) and Buzacott and Shanthikumar(1993).

Approximate Methods

Approximate methods are classified into simulation and decomposition models (Osorio and Berlaire, 2009). Simulation models are applied when the system has a high level of complexity or there are no analytical methods for calculating its properties (Law and Kelton, 2000).

Decomposition

There is an abundance of literature on the Markov case, particularly the papers of Takahashi, Miyahara and Hasegawa (1980) for Markov network systems, Perros and Altiok (1986), Altiok and Perros (1987), Gershwin (1987), Kerbache and MacGregor Smith(1987), Akyildiz (1988) and Hopp and Spearman (2000). Some of these procedures are described in Balsamo, Nitto-Persone and Onvural (2001). The approximation proposed by Takahashi, Miyahara and Hasegawa (1980) is explained below.

Approximation of Takahashi, Miyahara, and Hasegawa(1980)

This approximation for Markov systems is simple to implement and will be covered in the rest of the chapter.

Figure 6. Two stations with a buffer in front of station i+1

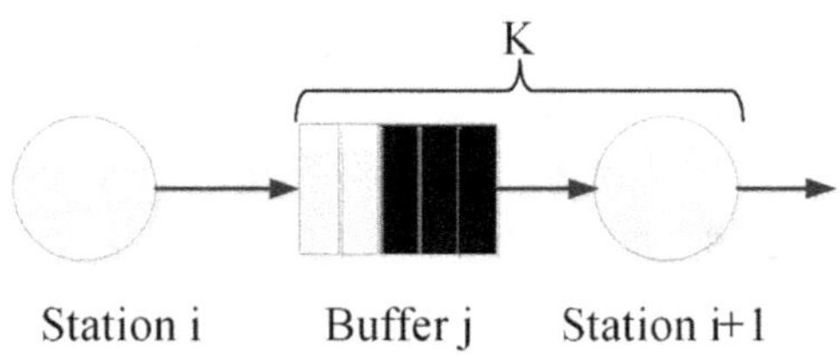

In Figure 6 there are two stations. Station i has enough space for the pieces that line up before it. Furthermore, there is a buffer before station $i+1$ with b_j available spaces for the line of pieces to be processed, therefore the capacity of any station is $K_i = b_j+1$. Each station has room to process one piece at a time.

When the piece or customer enters station i the service process starts, as soon as the service process finishes, the customer checks to see if at least one place is available in the next station. One of the following cases may arise:

If there is enough space in the next stage, then they leave the station making the space available for the following piece. Whereas if there is no room, then the piece or customer stay in the station for longer (there is a blockage). They will stay there until they receive the signal that there is at least one place available in the next station and only then will they be able to leave that stage.

Let the average residence time ($t_{r,i}$) be the average time a station is occupied by a piece or customer, which is the sum of the average process time that the piece requires ($t_{s,i}$) plus the average time that the next station has all its spaces occupied ($t_{K,i+1}$) (Buzacott and Shantikumhar, 1993):

$$t_{r,i} = t_{s,i} + t_{K,i+1} \tag{1}$$

Observe in (1) that if there is at least one place in station $i+1$ then $t_{r,i} = t_{s,i}$ and the piece will only occupy station i for the period corresponding to the service.

From the perspective of machine i, in a long term operation, station $i+1$ will have periods when it has space to accept customers and periods when the buffer will be full, blocking its predecessor (Figure 7).

From the perspective of station i, the length of the blocking periods will happen randomly. The average time that a station remains blocked is:

$$t_{i,K} = p_{K,i} t_{r,i} \tag{2}$$

Figure 7. Blocking in a long term operation

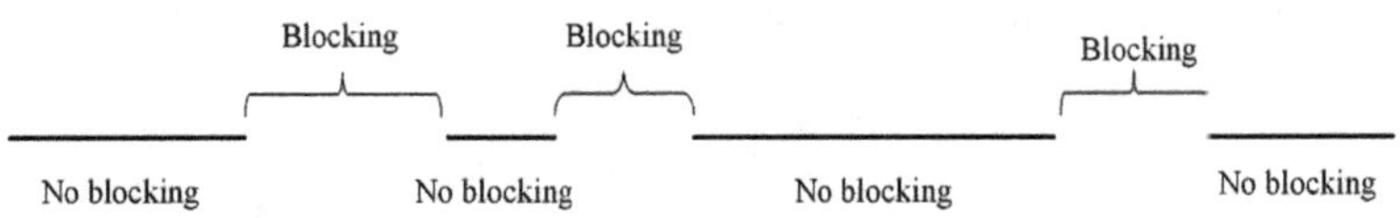

In (2) $p_{K,i}$ is the probability of all the places being occupied (Takahashi, Hasegawa and Miyahara, 1980). Substituting in the first equation gives us:

$$t_{r,i} = t_{s,i} + p_{K,i+1} t_{r,i+1} \tag{3}$$

The average time that all the spaces in station $i+1$ are occupied is $p_{K,i+1} t_{r,i+1}$. In the case of a system with $M/M/1/K$ stations, then $p_{K,i}$ is calculated using the following expression:

$$p_{K,i} = \frac{\rho_i^{K_i}\left(1-\rho_i\right)}{\left(1-\rho_i^{K_i+1}\right)} \tag{4}$$

Substituting (4) in (3):

$$t_{r,i} = t_{s,i} + \frac{\rho_{i+1}^{K_{i+1}}\left(1-\rho_{i+1}\right)}{\left(1-\rho_{i+1}^{K_i+1}\right)} t_{r,i+1} \tag{5}$$

where $\rho_{i+1} = \lambda_{e,i} t_{r,i+1}$. It must be remembered that the output flow in $M/M/1/K$ stations is denoted as $\lambda_{e,i}$, in a system in series the output from station i becomes the entry flow for station $i+1$ as we have explained in chapter 5 on M/M/1 systems in series.

Now expression 5 will be applied recursively to the system in Figure 8. Let us assume that the stations are $M/M/1/K$.

Starting with the last station, Station 3 is the last of the line and is not blocked, so the average residence time is equal to the average service time, in other words $t_{r3} = t_{s3}$. Continuing on with station 2 we have (6):

Figure 8. A system with three stations and two buffers

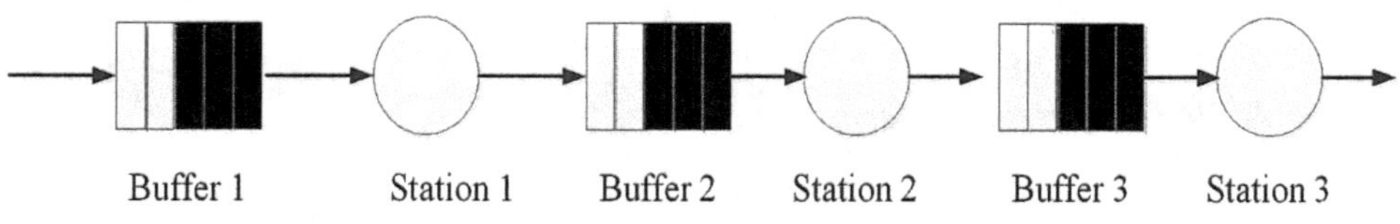

$$t_{r,2} = t_{s,2} + \frac{\rho_3^{K_3}\left(1-\rho_3\right)}{\left(1-\rho_3^{K_3}\right)} t_{r,3} \tag{6}$$

where $\rho_3 = \lambda_{e,2}\, t_{r,3}$. Finally for station 1 we have (7):

$$t_{r,1} = t_{s,1} + \frac{\rho_2^{K_2}\left(1-\rho_2\right)}{\left(1-\rho_2^{K_2}\right)} t_{r,2} \tag{7}$$

The congestion at station 2 is $\rho_2 = \lambda_{e,1}\, t_{r,2}$, while in the case of station 1, the congestion is $\rho_1 = \lambda_{e,1} t_{r,1}$ and the effective output of station i is: $\lambda_{e,i} = (1-p_{k,i})\, \lambda_{e,i-1}$ and is obtained once $p_{k,i}$ is known. The recursive equation is used in the following algorithm for calculating the properties of a system with buffers:

```
Reading parameters→n, t_{s,i}, K_i, λ_1
End = False
t_{r,i}= t_{s,i} for i=n, …,1
λ_1 →λ_{e,i} for i=n, …,1
Calculate: ρ_i
Resolve the flow equations
```

$$\lambda_i = \gamma_i + \sum_{j=}^{n} \lambda_j p_{ji}; i = 1,2,\ldots,n$$

```
where p_{ji} is the probability of the customer going from station
j to station i
For i= n,…,1
Calculate p_{K,i} with (5)
Calculate t_{r,i} = t_{s,i} + p_{K,i} t_{ri+1}
Update the output flow λ_{e,i}= (1-p_{K,i})λ_{e,i-1} i=n,…, 1
1→ iterations
        For i= n,…, 1
        Calculate ρ_i
        Calculate p_{K,i} with (5)
        Calculate t_{r,i} = t_{s,i} + p_{k,i} t_{ri+1}
        Update the output flow λ_{e,i}= (1-p_{k,i})λ_{e,i-1}
Iterations = iterations + 1
Until End = true
Calculate WIP_i, CT_i, i=n,…,1
```

In order to get $p_{k,i}$ we need to determine ρ_i and, in turn, $t_{r,i}$. In the first iteration we assume $\lambda_1 = \lambda_{e,i}$ for all the stations. Then its properties are determined by updating $\lambda_{e,i}$ for each station from station n and backwards to station 1, repeating this process until the maximum number of iterations is reached or until the difference in the value of $t_{r,i}$ in two consecutive iterations is less than a particular threshold.

The above procedure can be applied not only to systems in series but also to systems in network, we only need to get the entry flows to each station.

Example 6.1: The average service time and capacity of three stations are given in the Table 1.

Demand is 4.0 customers a minute. Let us assume that the inter-arrival time and service time follow an exponential distribution.

Table 2 gives the results obtained by applying the procedure we explained above and compares them with simulation.

The total cycle time is 2.789 minutes and the work in process is 7.77 units. Given that the stations have the same service time, the production will be that of the last station, in this case, 2.78 pieces

Example 6.2: Let us now assume that the service times are as follows: t_s= [1/5, 1/6, 1/5]. The results are shown in Table 3.

Table 1. Service time and capacity

Station	1	2	3
$t_{s,i}$**(min)**	1/3	1/3	1/3
K_i	5	5	5

Table 2. Performance of the system

Station	**Approximation**		**Simulated**	
	WIP$_i$	*CTi*	*WIP*	*CT*
1	2.84	1.02	3.7	1.45
2	2.65	0.954	3.17	1.26
3	2.28	0.819	2.2	0.97

Table 3. Performance of the system

Station	Approximation		Simulated	
	WIP_i	CTi	WIP	CT
1	1.74	0.47	2.09	0.56
2	1.48	0.39	1.64	0.43
3	1.63	0.44	2.14	0.57

The average cycle time is 1.3 minutes (1.42 minutes, simulated) and the amount of work in process is 4.85 units (5.9 units, simulated). In this case the system's output is 3.64 pieces (Table 3).

Example 6.3: Let us consider a system like the one shown in Figure 9. The data for the service time (in hours) are t_s= [1/7, 1/4, 1/3]. The entry flow is 4.5 pieces an hour. The first station has space for 5 pieces, while station 2 has the capacity for 6 pieces and station 3 has the capacity for 7 pieces. Let us assume that the service time and inter-arrival time follow an exponential distribution.

The cycle time in the network is 1.03 hrs. (0.94 hrs, simulated) and the total work in process is 4.53 units (4.17 units, simulated) (Table 4).

Example 6.4: Let us assume that we have a production line with the characteristics shown in Table 5.

Let us assume that 2 pieces arrive in the system every minute. Let us assume that the time between failures, the repair time, the service time and

Figure 9. Network of example 3

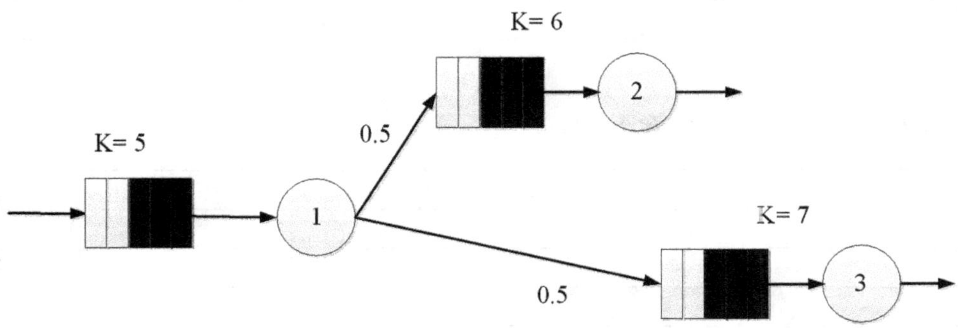

Table 4. Results of example 3

Station	Approximation		Simulated	
	WIP_i	CTi	WIP	CT
1	1.44	0.33	1.28	0.30
2	1.13	0.50	1.03	0.48
3	1.96	0.90	1.8	0.70

the time between consecutive customers follow an exponential distribution (Table 5). The results are in Table 6.

Although the decision to set limits on the amount of work in process is sometimes unavoidable, this has the effect of lowering the productivity of a production line and increasing the probability of a station being blocked. Both of which are negative effects. Observe that as the available space is limited, the number of customers entering the system is also is limited, therefore λ-λ_e, which is the number of customers that are not processed in a long-term operation, increases.

Furthermore, having low inventory limits ensures low storage costs (think, for example, of perishable goods), the cycle time will also be low and makes it easier for the worker to keep a certain order in their work area.

Table 5. Data of example 4

Station	1	2	3
t_s (min)	1/3	¼	1/3
K	4	5	4
MTTR	500	300	400
MTTF	60	60	60

Table 6. Results of example 4

Station	WIP_i	CTi
1	1.31	0.73
2	1.33	0.69
3	1.28	0.72

A Comment on Productivity: The Bowl Phenomenom

The design and planning of a production system are particularly important tasks for specialists in the study of production systems. In an environment with a high level of competition between companies, any improvements that would increase productivity are more than welcome.

A common practice in industry is to operate the production lines by allocating the same workload or congestion to each station, in other words, endeavoring to balance the system's workload as this is considered to be the most efficient way to operate a production line (Figure 10). The studies on productivity led to the Assembly-Line Balancing Problem (Lopez, 2014).

Having a perfectly balanced line is an ideal, as production systems are subject to variability, owing to the random nature of demand, service time and system failures.

During their research into the problem of allocating the workload to production systems with variable service times, Hillier and Bolling (1966) discovered that systems with a balanced workload are less efficient (produce less) than lines with an unbalanced workload that alternate between slow stations and fast stations.

Assuming that the service times follow an exponential probability distribution, Hillier and Bolling (1966) concluded that a strategy for increasing the throughput (Th) of the production line would consist of creating an

Figure 10. A balanced production line, all stations with the same congestion

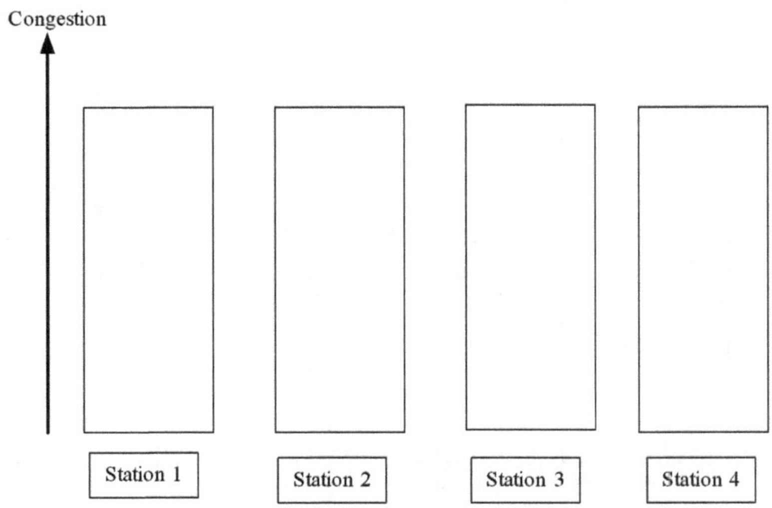

imbalance by making the machines at each end have a higher workload (work faster) that the ones in the center (Figure 11).

Plotting the workload of the stations gives us a pattern shaped like a bowl, where the service times of the stations in the central area are shorter than the service times corresponding to the stations at the two ends (Hillier and Bolling, 1966; Hillier and So, 1991, 1993; Castellucci and Costa, 2014).

This result has created controversy with a variety of relevant studies having been carried out, though no definitive conclusion has been reached yet. This controversy can be summed up by the following arguments (Castellucci and Costa, 2015, da Silva; Hillier, So and Boling, 1993; Hillier and So, 1996; Hillier, 2013; Sampaio and Passos, 2016):

1. The conclusions drawn by Hillier and Bolling were obtained by experimenting on 4-station lines (Hillier and Bolling, 1966) when, in reality, production lines entail a lot more stages.
2. The experiments do not enable us to draw any conclusion as to whether the improvement is the result of unbalancing service times or because of variability in service times.
3. The experiments assumed that the stations are not subject to failures.
4. The conclusions were reached by experimenting with systems that are of no interest or else are not commonly found in industry: high variability and only a few stations (Smut and Perkins, 1985)

Figure 11. The bowl phenomenon: Stations 2 and 3 are faster than stations 1 and 4.

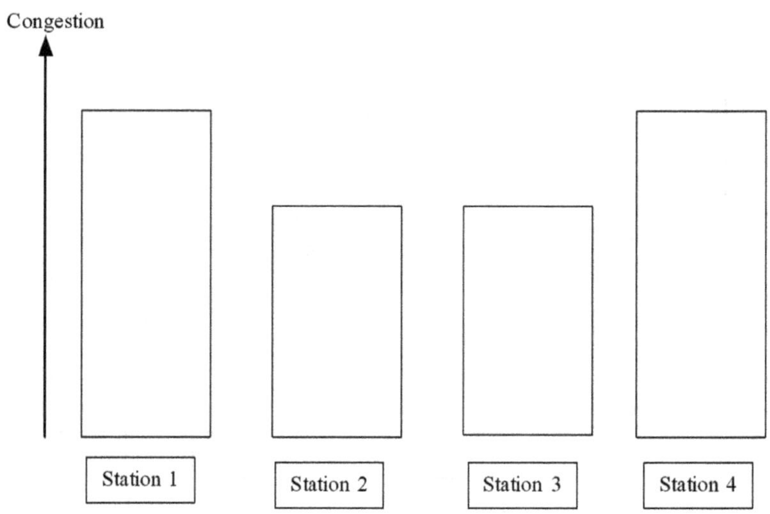

Example 6.5: Let us assume that there is a production line with the characteristics shown in Table 7.

All the stations have the same service rate (it is balanced) and have limited queueing capacity (Table 7). The stations are Markov stations. By applying the procedure provided by Takahashi, Miyahara and Hasegawa (1980), we get the results given in Table 8.

Th = 0.869 is observed in the balanced case; however by making station 2 work faster we observe a maximum value of *Th* = 0.8924 when $\mu_2 = 3$, whereas if station 3 is the one that works faster we get a maximum value of *Th* = 0.8914 when $\mu_3 = 3$ (Figure 12).

The rise in the value of *Th* is known to be in the 1 – 2% range. More recent studies show that even when there are deterministic service times, it is a good idea to take the option of unbalanced production systems into account. On the other hand, later studies have found that when the work in process entails a high cost, the optimum allocation of workload in the stations does not follow a bowl pattern (Shaaban, McNamara, Hudson, 2014).

Table 7. Service time and K of stations

Service Rate	Capacity (*K*)
1.1	5
1.1	5
1.1	5

Table 8. Performance of a balanced line vs. an unbalanced line

Station	Balanced $\mu_1=\mu_2=\mu_3=1.1$		Station 2 Faster $\mu_2^*=2$		Station 3 Faster $\mu_3^*=2$	
	WIP_i	*CTi*	WIP_i	*CTi*	WIP_i	*CTi*
1	2.15	2.47	1.88	2.16	2.07	2.37
2	2.06	2.37	1.10	1.26	1.86	2.13
3	1.83	2.11	1.83	2.11	0.88	0.89
	Th = 0.869		*Th* = 0.8924		*Th* = 0.8914	

Figure 12. System throughput

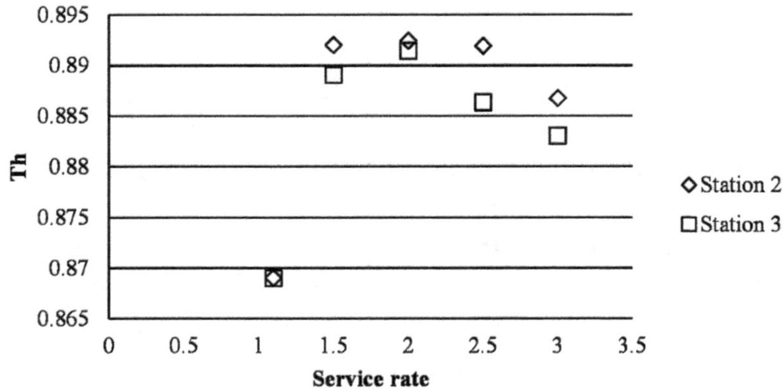

Single-Stage M/G/1/K Systems

The lack of equations for the analysis of systems with service times that do not follow an exponential probability distribution is made up for by using simulation as a tool.

Given the fact that these systems do not fulfill the lack-of-memory assumption, the modeling of this class of queues is complicated but several approximations have been developed for the blocking probability on which basis we can get the amount of work in process and cycle time. MacGregor Smith (2004) gives a list of some of these contributions. From these studies it has been concluded that the most robust approximation has been created by Erol Gelenbe (Gelenbe, 1975; MacGregor Smith 2004) and comparisons were made between this expression and the new approximation developed by MacGregor Smith.

Broadly speaking, the argument involved in obtaining the new approximation is as follows: A formula is derived from the p_K expression for *M/M/*1/*K* systems and the approximations provided by Tjims and Kimura, MacGregor Smith (2004) for the blocking probability of M/G/1/K systems based on the p_K expression of a *M/M/*1/*K* system:

$$p_K = \frac{(1-\rho)\rho^K}{1-\rho^{K+1}} \tag{8}$$

In expression (8) we relax the requirement for *K* to be an integer number in order to get an expression (9) for capacity as a function of congestion:

$$K = \frac{\ln\left(p_K / \left(1-\rho+p_K\right)\right)}{\ln\left(\rho\right)} \tag{9}$$

On the other hand, Kimura developed the following expression for estimating buffer size:

$$\tilde{B}_{\varepsilon}\left(C^2\right) = B_{\varepsilon}\left(M\right) + NINT\left[\frac{1}{2}\left(C^2-1\right)\sqrt{\rho}B_{\varepsilon}\left(M\right)\right] \tag{10}$$

where B(M) in equation (10) is the expression for the size of buffer in a Markov system and NINT indicates the need to round it up or down to the nearest integer. By substituting the K expression we get the following equation (11) for the blocking probability in M/G/1/K systems (MacGregor Smith, 2004, 2011a):

$$p_K = \frac{\rho^{\beta_1}\left(\rho-1\right)}{\rho^{\beta_2}-1} \tag{11}$$

with β_1 and β_2 being:

$$\beta_1 = \frac{C_S^2\sqrt{\rho e^{-C_S^2}} - \sqrt{\rho e^{-C_S^2}} + 2K}{2 + C_S^2\sqrt{\rho e^{-C_S^2}} - \sqrt{\rho e^{-C_S^2}}} \tag{12}$$

$$\beta_2 = 2\frac{C_S^2\sqrt{\rho e^{-C_S^2}} - \sqrt{\rho e^{-C_S^2}} + K + 1}{2 + C_S^2\sqrt{\rho e^{-C_S^2}} - \sqrt{\rho e^{-C_S^2}}} \tag{13}$$

where $C^2{}_S$ is the squared coefficient of variation for the service time and e is Euler's number (12, 13). In order to get p_0 we resort to the following expression:

$$\lambda\left(1-p_K\right) = \mu\left(1-p_0\right) \tag{14}$$

Substituting the terms in (14) that are already known and finding p_0:

$$p_0 = \frac{(\rho - 1)}{\rho^{2\left(\frac{C_S^2\sqrt{\rho}-\sqrt{\rho}+K+1}{2+C_S^2\sqrt{\rho}-\sqrt{\rho}}\right)} - 1} \tag{15}$$

The probability distribution of the number of customers in the system is calculated using the formula for M/M/1/K systems and (15):

$$p_n = p_0\rho^n \tag{16}$$

Figure 13 shows the blocking probability of an M/G/1/K system using MacGregor Smith's approximation for the case of exponential service time ($C^2_S = 1$) as a function of ρ and different values of K; in the exponential case, the approximation gives the same result as the exact expression.

Figure 14 shows the profile of the blocking probability for K = 5 and C^2_S =0.5, 1 and 4 respectively.

The following program builds the blocking probability graph from MacGregor Smith's expression.

Figure 13. p_K approximation, $C^2_S = 1$

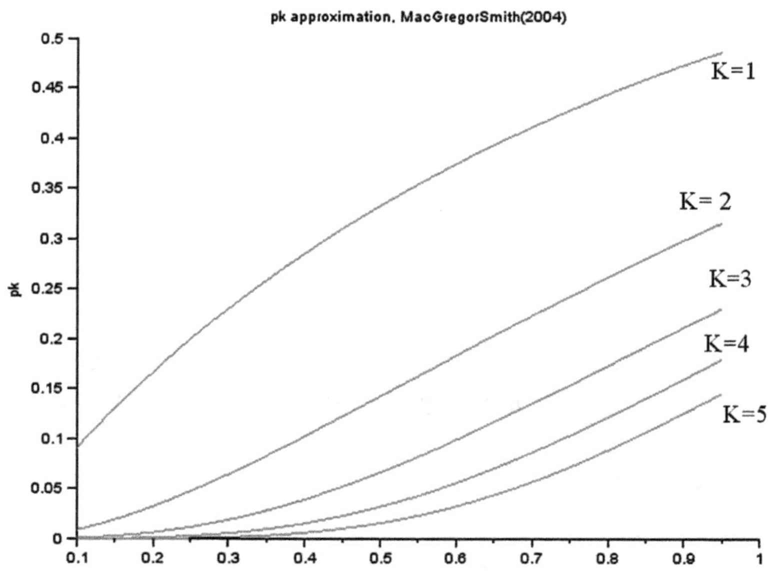

Figure 14. p_K approximation, K = 5

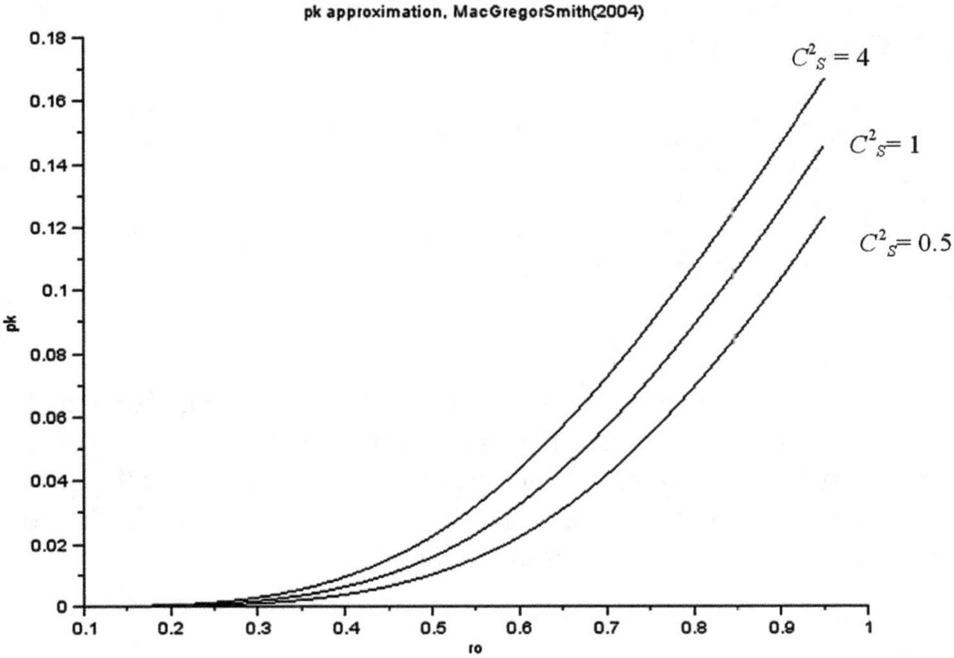


```
clf()
x=[0.1:0.01:0.95]';
xtitle("pk approximation, MacGregor Smith(2004)")
ylabel("pk")
xlabel("ro")
c2s=1.0;
K=5;
plot2d(x, x.^((c2s*sqrt(x./exp(c2s))-sqrt(x./exp(c2s))+K*2)./
(2+c2s*sqrt(x./exp(c2s))-sqrt(x./exp(c2s)))).*(x-1)./
(x.^(2*(c2s*sqrt(x./exp(c2s))-sqrt(x./exp(c2s))+1+K)./
(2+c2s*sqrt(x./exp(c2s))-sqrt(x./exp(c2s))))-1), style=1);
```

The following program, also in Scilab, evaluates p_k:

```
function [pblok]=pk(x, K, c2s)
 y=x.^((c2s*sqrt(x./exp(c2s))-sqrt(x./exp(c2s))+K*2)./
(2+c2s*sqrt(x./exp(c2s))-sqrt(x./exp(c2s)))).*(x-1)./
(x.^(2*(c2s*sqrt(x./exp(c2s))-sqrt(x./exp(c2s))+1-K)./
(2+c2s*sqrt(x./exp(c2s))-sqrt(x./exp(c2s))))-1)
 [pblok]=y
endfunction
```

From the approximation of the blocking probability, MacGregor Smith derives the following formula for the work in process of an M/G/1/K station (MacGregor Smith 2004, 2011a):

$$WIP = \begin{cases} \dfrac{\rho\left(\ln\left(1/\rho\right) + p_K \ln\left(p_K / \left(1-\rho+p_K\right)\right) + p_K \ln\left(\rho\right)\right)\left(2 + C_S^2\sqrt{\rho e^{-C_S^2}} - \sqrt{\rho e^{-C_S^2}}\right)}{2\ln\left(\rho\right)\left(\rho-1\right)} & C_S^2 \le 1 \\ \dfrac{\rho\left(\ln\left(1/\rho\right) + p_K \ln\left(p_K / \left(1-\rho+p_K\right)\right) + p_K \ln\left(\rho\right)\right)\left(2 + C_S^2\sqrt{\rho e^{-1}} - \sqrt{\rho e^{-}}\right)}{2\ln\left(\rho\right)\left(\rho-1\right)} & C_S^2 > 1 \end{cases} \quad (17)$$

The relevance of expressions (11), (12), (13), (14), (15), (16) and (17) lies in the fact that there is an analytical means for quantifying the blocking probability and the amount of work in process in a queue where the service time does not follow an exponential distribution. In order to assess the work in process, we can use the following program:

```
x=[0.1:0.1:0.9];
c2s=1.0;
K=5;
if (c2s<=1) then
 wipp=x.*((log(1./x))+pk.*log(pk./(1-x+pk.*x))+pk.*log
(x)).*((2+c2s*sqrt(x./exp(c2s))-sqrt(x./exp(c2s)))))./
((2*log(x).*(x-1)));
else
 wipp=x.*((log(1./x))+pk.*log(pk./(1-x+pk.*x))+pk.*log(x)).*((2
+c2s*sqrt(x./exp(1))-sqrt(x./exp(1))))./((2*log(x).*(x-1)));
end
[wip]=[wipp];
```

The following program in Scilab plots the work in process for this case using $C^2_S \geq 1$ and K=5 (Figure 15).

```
clf()
x=[0.1:0.1:0.9];
xtitle("WIP approximation, MacGregor Smith(2004)")
ylabel("WIP")
xlabel("ro")
c2s=1.0;
K=5;
plot2d(x, x.*((log(1./x))+pk.*log(pk./(1-x+pk.*x))+pk.*l
og(x)).*((2+c2s*sqrt(x./exp(c2s))-sqrt(x./exp(c2s)))))./
((2*log(x).*(x-1))), style=1);
```

Figure 15. WIP vs ρ, K= 5

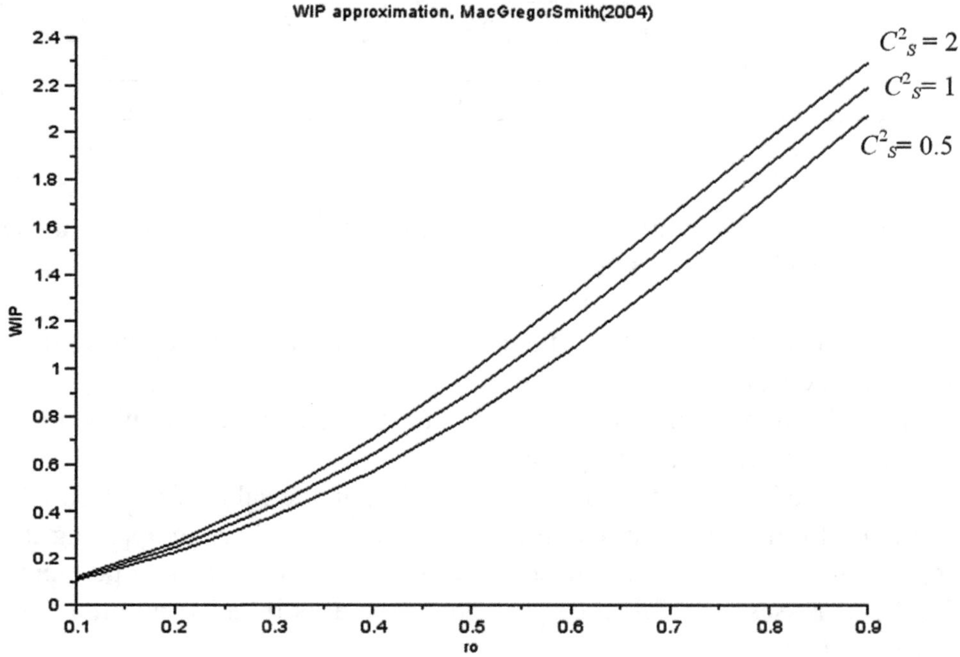

Example 6.6: Let us assume that there is a work station with a single machine. A piece arrives at the station every two minutes and the inter-arrival time follows an exponential probability distribution. The station's average service time is 2.25 minutes per piece with a squared coefficient of variation of 1.5. Given the size of the pieces being processed, there is sufficient space in front of station for 5 pieces to queue. Get the performance measures for this work station.

The blocking probability is $p_K = 0.23$. The probability of the system being empty is $p_0 = 0.13$. The congestion is $\rho = 1.125$. The probability of finding $n = 1, 2, 3, 4$ and 5 customers in the station will have a cumulative probability that is calculated using $1 - p_K - p_0$. For this example, we have 1-0.23-0.13=0.65 (MacGregor Smith, 2011a).

The amount of work in process will be 2.98 pieces. The throughput will be $0.5(1 - 0.2) = 0.385$ and the average cycle time is $2.98/0.4 = 7.74$ minutes.

Example 6.7: Let us assume that the person in charge of the station implements a project to lessen the variance of the service time. This project implies

Table 9. Results of example 7

Month	C^2_S	WIP	Th	CT
1	1.0	2.84	0.39	7.64
2	0.75	2.80	0.395	7.07

training the workers and making improvements to the devices of the machine in order to speed up the operation.

On drawing up the work plan the improvements are expected to result in a drop in the value of C^2_S.

Apply the approximations to estimate the changes in the station's work in process, cycle time and throughput.

This Table 9 shows the estimates for the drop in the value of C^2_S, followed by the performance measures obtained for each case through applying the approximations. Taking the current state as a reference, in the second month the Th increases by 2.5%, the cycle time decreases by 8.7% and the WIP decreases by 6.4%.

M/G/C/K Systems

The development of blocking probability approximations gets ever more complicated as the number of the system's servers increases. MacGregor Smith (2003) has developed expressions for up to 5 servers.

The process for obtaining an analytical expression for the blocking probability of *M/G/c/K* systems is similar to the case where there is a single server. The equations for each case respectively are:

$$p_K = \frac{1}{c^{K-c}c!}\left(\frac{\lambda}{\rho}\right)^K p_0 \tag{18}$$

$$p_0 = \left\{\sum_{l=0}^{c-1}\frac{1}{l!}\left(\frac{\lambda}{\mu}\right)^l + \sum_{l=c}^{K}\frac{\left(\lambda/\mu\right)^l}{c!c^{l-c}}\right\}^{-1} \tag{19}$$

where the condition $\lambda/(c\mu) \neq 1$ in (18) and (19) has to be fulfilled. In order to obtain a p_K expression, we need to first set the value of the number of servers c. For example in the $M/G/2/K$ case, the expression for the value of K is:

$$K = -\frac{\ln\left(\frac{1}{2}\left(p_K\left(2\mu+\lambda\right)\right)/\left(2\mu-\lambda+\lambda p_K\right)\right)}{\ln\left(2\frac{\mu}{\lambda}\right)} \tag{20}$$

By substituting the equation (20) in Gelenbe's formula, the following expression is obtained for the blocking probability:

$$p_K = \frac{2\rho^{2\beta_1}\left(2\mu-\lambda\right)}{-2\rho^{2\beta_2}\lambda+2\mu+\lambda} \tag{21}$$

In this case β_1 and β_2 in equation (21) respectively are:

Figure 16. P_k vs congestion of an M/G/2/K system

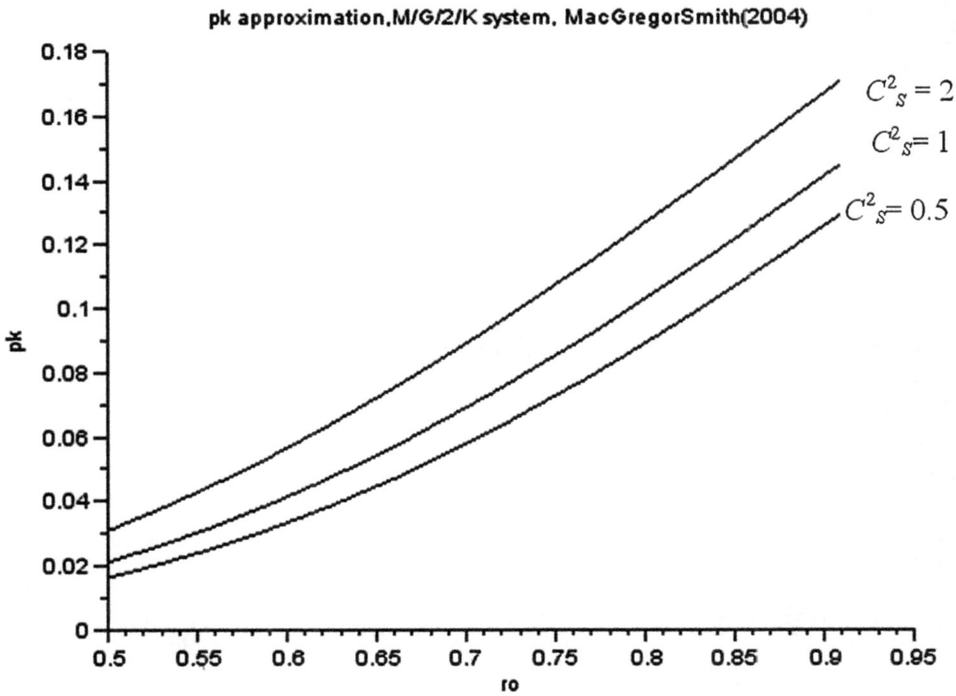

$$\beta_1 = \beta_2 = \frac{2 + C_S^2\sqrt{\rho/_e} - \sqrt{\rho/_e} + B}{2 + C_S^2\sqrt{\rho/_e} - \sqrt{\rho/_e}} \tag{22}$$

where B in (22) is the size of the buffer, in other words, the number of spaces available in the queue for the customers to line up.

The following Scilab code calculates the blocking probability and the probability of the system being empty p_0. The last section evaluates the probability distribution of having n customers in the system for the case of a *M/G/2/K* queue and the calculations are made within the $\rho = (0.5 - 0.9)$ range. Figure 16 gives the results for three different values of C^2_S.

```
mu=[1.1:0.1:2]';
lambda=2;
K=4;
c2s=0.5;
c=2
x=lambda./(c*mu);
beta1=2+c2s*sqrt(x./exp(1))-sqrt(x./exp(1))+(K-1);
beta2=2+c2s*sqrt(x./exp(1))-sqrt(x./exp(1));
y=(2*x.^(2*beta1./beta2)).*(2*mu-lambda)./
(-2*lambda*x.^(2*beta1./beta2)+2*mu+lambda)
[pblock]=y;
suma=0;
l=0;
while l<c
 suma=suma+(1/factorial(l))*(lambda./mu).^l;
 l=l+1;
end
sumaa=0.0
for l=c:K
 sumaa=sumaa+(lambda./mu)^l/((factorial(c))*c^(l-c))
end
p0=(suma+sumaa)^-1
xtitle("pk approximation,M/G/2/K system, MacGregor
Smith(2004)")
ylabel("pk")
xlabel("ro")
plot2d(x,y, style=1)
for l=1:c-1
 pn(l)=p0(1)*(lambda^l/mu(1)^l)/factorial(l)
end
for l=c:K-1
 pn(l)=p0(1)*(lambda^l/mu(1)^l)/(factorial(c)*c^(l-c))
end
```

For the *M/G/3/K* system:

$$p_K = \frac{-9\rho^{\beta_1}\mu\left(-3\mu+\lambda\right)}{-9\rho^{\beta_2}\mu\lambda+6\mu^2+4\lambda\mu+\lambda^2} \tag{23}$$

where β_1 and β_2 in equation (23) respectively are:

$$\beta_1 = \beta_2 = \frac{6+3C_S^2\sqrt{\rho/e}-3\sqrt{\rho/e}+2B}{2+C_S^2\sqrt{2+\rho/e}-\sqrt{\rho/e}} \tag{24}$$

Given that the p_K and p_0 expressions (23 and 24)are approximations and give the extreme values, then the remaining p_n should be adjusted to the difference $1 - p_K - p_0$. In MacGregor Smith(2003) the expressions of c= 4 and 5 servers are also developed.

In order to calculate the amount of work in process, we apply the general work-in-process equation (25):

$$WIP = \sum_{n=1}^{K} np_n \tag{25}$$

Example 6.8: Let us look again at example 6 and assume that instead of making improvements to lower the value of C2S we decide instead to add a machine to the station, so we will now have two servers in parallel and let us assume that there is no difference between the machines. Carry out an analysis of the system with these conditions and compare the results obtained.

By applying the corresponding expressions, we get p_K= 0.0136 and p0= 0.2865. The value of Th= 0.4935. The WIP= 1.455 pieces and the cycle time is 2.94 minutes.

Example 6.9: Let us assume that a work station processes pieces at a service rate of 15 an hour, the squared coefficient of variation of the service time is 2 and the station has two machines in parallel. Demand is 17 pieces per hour, while the inter-arrival times follow an exponential distribution. Let us assume that the station has capacity for 6 spaces: 4 lined up and 2 in the servers. Characterize the work station.

The blocking probabilities and the probabilities of the system being empty are 0.018 and 0.28 respectively.

The work in process is 1.5 units and the throughput, 16.69 while the cycle time is 0.089 hours or approximately 5.94 minutes.

Network Systems With M/G/1/K Stations

With the approximations shown in the above section, it is feasible to analyze more complex systems. The details of the implemention of the expressions were given in the General Expansion Method proposed by Kerbache and MacGregor Smith(1987) for modeling networks *M/G/1/K*: The buffer allocation problem (MacGregor Smith and Cruz, 2005; Cruz, Duarte and van Woensel, 2008), traffic flow (van Woensel and Vandaele,2006) and throughput maximization (Cruz, Kendall, While, Duarte and Brito, 2012); in MacGregor Smith(2011), in particular, an analysis was made of the robustness of the GEM for M/G/1/K systems, confirming their reliability in the calculation of the throughput, cycle time and amount of work in process.

The GEM has been improved for the analysis of M/G/c/*K* systems (Van Woensel and Cruz, 2014, Cruz and van Woensel, 2015) and M/G/c/c systems applied to vehicular flow (Jain and MacGregor Smith, 1997) and the design of evacuation policies (Cruz and MacGregor Smith, 2007).

Takahashi's Method Adapted for Networks With M/G/1/K Stations

The procedure was adapted, as shown below, for the purpose of estimating the properties with M/G/1/K stations. The modifications consist of estimating the blocking probability p_0 and p_K using the approximations (11) - (16). The residence time $t_{r,I}$ is then determined. The work in process (WIP) for each station is estimated using the equation (25).

The following program assesses the properties of a network system with *M/G/1/K* stations.

```
function [x]=queueMG1K(n,gamm,c2g,P,ts,c2s,k)
 //Subroutine that evaluates properties of MM1K queueing nets
 //Uses the procedure published by takahashi(1980)
 // n: number of stations
 // gamm: external flow rate
 // c2g: square coefficient of variation of arrivals
 //P: Matrix probability
```

```
// ts: service time
// c2s: square coefficient of variation of service time
//k: capacity of the system
pt=P'
I=eye(n,n)
//Evaluation of lambda
lamb=inv(I-pt)*gamm
ro=lamb.*ts
// Probability of blockin. First Iteration
beta1=(sqrt(ro.*exp(-c2s)).*c2s-sqrt(ro.*exp(-c2s))+2*k)
beta1= beta1./(2+sqrt(ro.*exp(-c2s)).*c2s-sqrt(ro.*exp(-c2s)))
beta2=(1+k+sqrt(ro.*exp(-c2s)).*c2s-sqrt(ro.*exp(-c2s)))
beta2=beta2./(2+sqrt(ro.*exp(-c2s)).*c2s-sqrt(ro.*exp(-c2s)))
beta2=2*beta2
pk=ro.^(beta1)
pk=pk.*(ro-1)
pk=pk./(ro.^(beta2)-1)
lambe=lamb.*(1-pk)
th=0.0
for i=1:n
e(i)=i
for ii=n:-1:1
tr(ii)=0.0
sumatr=0.0
yy=0.0
for j=1:n
if (P(ii,j)>0) then
sumatr=sumatr+pk(j)*tr(j)
yy=yy+1
end
end
tr(ii)=ts(ii)+sumatr
end
roe=lambe.*tr
// pk with MacGregor Smith approximation
beta1=(sqrt(roe.*exp(-c2s)).*c2s-sqrt(roe.*exp(-c2s))+2*k)
beta1= beta1./(2+sqrt(roe.*exp(-c2s)).*c2s-sqrt(roe.*exp(-
c2s)))
beta2=(1+k+sqrt(roe.*exp(-c2s)).*c2s-sqrt(roe.*exp(-c2s)))
beta2=beta2./(2+sqrt(roe.*exp(-c2s)).*c2s-sqrt(roe.*exp(-
c2s)))
beta2=2*beta2
pk=roe.^(beta1)
pk=pk.*(roe-1)
pk=pk./(roe.^(beta2)-1)
p0=(roe-1)
p0= p0./(roe.^(beta2)-1)
end
// p0, WIPs, TCs
```

```
  for ii=1:n
  kmax=k(ii)
  for jj=1:kmax
  pn(ii,jj)=p0(ii)*roe(ii)^jj
  end
  wips(ii)=0.0
  for jj=1:kmax
  wips(ii)=wips(ii)+pn(ii,jj)*jj
  end
  end
  tcs=wips./lambe
  [x]=[e,roe,pk,p0,lambe,wips,tcs]
endfunction
```

Example 6.10: Let us consider a system with three stations. The customers enter through station 1 with a inter-arrival time of 1 minute. The average service time and capacity K at each station is as shown in Table 10.

The path follow by the customers is given in the Figure 17.

Table 11 shows the results of applying the modified method of Takahashi, et al. (1980) using p_K approximation.

Once the system's performance measures have been determined, the next step is to carry out the analysis needed to improve the indicators, in other

Table 10. Data of example 10

Station	1	2	3
t_s (min)	0.7	1.5	1
K	5	5	6

Figure 17. Network of example 10

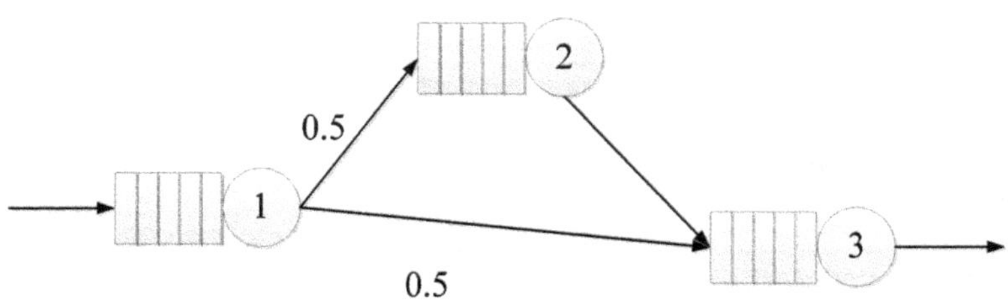

Table 11. Results of example 10

Station	M/G/1/K		
	WIP	*CT*	p_K
1	2.83	2.06	0.2179
2	1.60	1.68	0.063
3	2.28	1.63	0.137

words, a search is made for the conditions that would optimize the performance, subject to the constraints imposed by capacity, space, etc.

In queueing networks with a finite queue there are several optimization problems that have been studied for several decades; MacGregor Smith(2009) makes a classification of optimization models, which is set out below.

A General Decision Problem

In the general problem, the decision variables are K and c, which are integers, and μ and λ that are continuous variables (MacGregor Smith, 2010):

Maximize

$$G = \sum_{i=1}^{n} P_i \left(\lambda_i \left(1 - p_{K,i}\right)\right) - \sum_{i=1}^{n} C_{WIP,i} WIP_i \tag{26}$$

Subject to:

$$\sum_{i=1}^{n} K_i \le K_{Total} \tag{27}$$

$$\sum_{i=1}^{n} c_i \le Servers_{Total} \tag{28}$$

$$\sum_{i=1}^{n} \mu_i \le Capacity_{Total} \tag{29}$$

$$\sum_{i=1}^{n} \lambda_i \leq Flow_{Total} \tag{30}$$

$$K_i \geq 1 \text{ and integer} \tag{31}$$

$$c_i > 1 \text{ and integer,} \tag{32}$$

$$\mu_i > 0 \tag{33}$$

$$\lambda_i > 0. \tag{34}$$

where (26) is the total gain and is the difference in revenue minus the cost of the flow through a station. The total available space (27) is limited to a certain number of spaces; the number of servers must not exceed a certain number (28), the total service capacity is constrained (29) and, finally, the total flow of customers circulating through the network (30) is also limited. Constraints (31) and (32) correspond to the integer decision variables and (33) and (34) correspond to the continuous variables. This model is mixed integer nonlinear, with constraints of the type to be found in the knapsack problem, which is known as an NP-complete problem.

This model considers the most common factors that an administrator must take into account when making a decision; it is worth mentioning that other optimization problems derive from this general problem and these are briefly described below.

Problem of Minimizing the Total Waiting Cost

This problem has three variants, depending on the decision variable used (MacGregor Smith, 2010). In the first case, the decision variable is the service rate and continuous, the optimum μ value minimizes the waiting cost (or the flow cost) through the stations.

Minimize

$$\text{Cos}\,t = \sum_{i=1}^{n} C_{WIP,i} WIP_i \tag{35}$$

Subject to:

$$\sum_{i=1}^{n} \mu_i = Capacity_{Total}, \tag{36}$$

$\mu_i > 0.$

Expression (35) and (36) of this problem were briefly covered in chapter 3 where the economic bottleneck was identified. In the second variant, the decision variable is the number of servers and an integer variable:

Minimize

$$\text{Cos}\,t = \sum_{i=1}^{n} C_{WIP,i} WIP_i \tag{37}$$

Subject to:

$$\sum_{i=1}^{n} c_i = Servers_{Total}, \tag{38}$$

$c_i > 1$ and integer.

Expression (37) is again the cost of waiting, equation (38) is the restriction of capacity. Finally, in the third variant, it is assumed that the decision-maker has control over the arrivals to the system (40). In this case the decision variable is λ and continuous:

Minimize

$$\text{Cos}\,t = \sum_{i=1}^{n} C_{WIP,i} WIP_i \tag{39}$$

Subject to:

$$\sum_{i=1}^{n} \lambda_i = Flow_{Total}, \tag{40}$$

$\lambda_i > 0$.

Optimum Buffer Size

In this case the decision variable is the K_i capacity of the stations; the variable is an integer and the problem as a whole corresponds to an integer nonlinear model.

Maximize

$$P = \sum_{i=1}^{n} P_i \left(\lambda_i \left(1 - p_{K,i} \right) \right) \tag{41}$$

Subject to:

$$\sum_{i=1}^{n} K_i = K_{Total}, \tag{42}$$

$K_i \geq 1$ and integer.

This problem is known as the Buffer Allocation Problem and the aim is to determine the place and number of allocated spaces in front of each station in order to maximize *Th* (equation 41) and the constraint (42) corresponds to the number of spaces available for this purpose. This problem has been widely dealt with in the specialist literature. Other variants are minimizing the *WIP* and minimizing the number of spaces allocated all along the production line.

One thing that complicates finding a solution to this problem is the fact that there is no expression for directly estimating the value of *Th* for a *K* vector; as we have already seen, an iterative process needs to be performed in order to be able to estimate said value before searching for the direction of improvement.

Figure 18 shows the behavior of the *Th* in respect of *K* and the congestion in a single station. It was created by relaxing the requirement for variable *K*

Figure 18. Th vs K = (1 – 7), ρ = (0.2 – 0.9), C^2_S=1, λ = 1

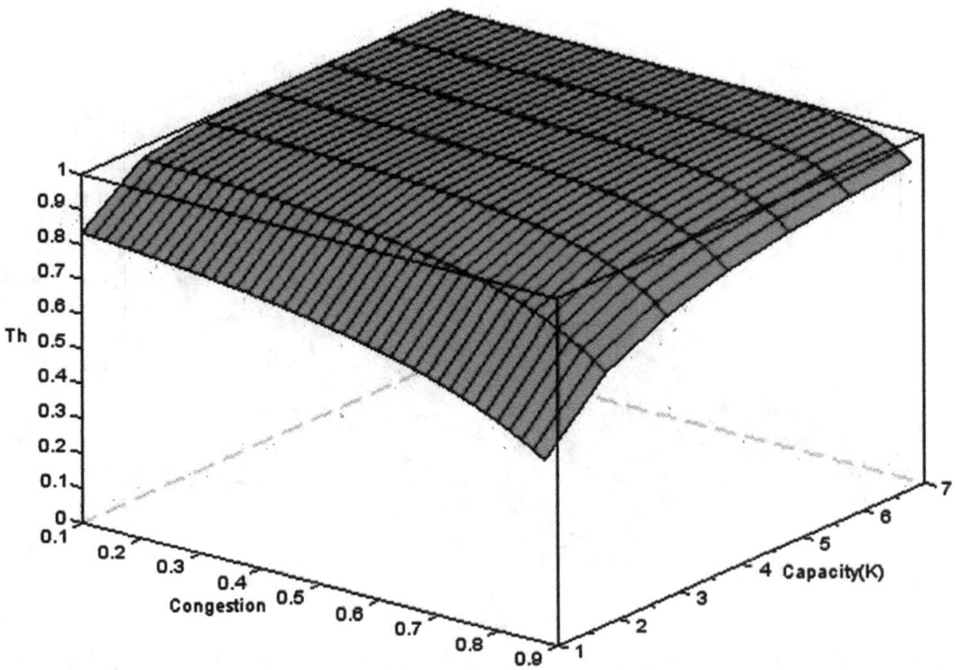

to be an integer. The graph shows a concave surface and we can see that, as the number of spaces increases, the value of *Th* converges on a limit value, in other *words* when $K \rightarrow \infty$, the station will start to behave like an *M*/*G*/1 system. Since the surface is constructed by making use of an approximation, consequently the value of *Th* is also approximate.

When there is a high congestion value and the buffer is very restricted, *Th* will be low, because the faster the station works (the value of $\rho\downarrow$) the greater the increase in the *Th*.

The search for the optimum operating conditions, taking *K* and ρ at the same time, is the equivalent of finding a direction of improvement that starts from the intersection where the value of *K* is low and the congestion level is high and moving towards the center of the figure.

Getting the analytical derivative of the expression of *Th* would be somewhat complicated owing to the structure of the p_K expression, however one alternative is to make an estimate of the rate of change of the *Th* in respect of *K*.

Figure 19 shows the rate of change of the *Th* for *K*=1, 2, 3 and 4 and assuming that *K* is continuous. In all the cases, there is an additional benefit to increasing the space in the buffer. However as capacity increases, the

Figure 19. ΔTh/ΔK of a single M/G/1/K queue

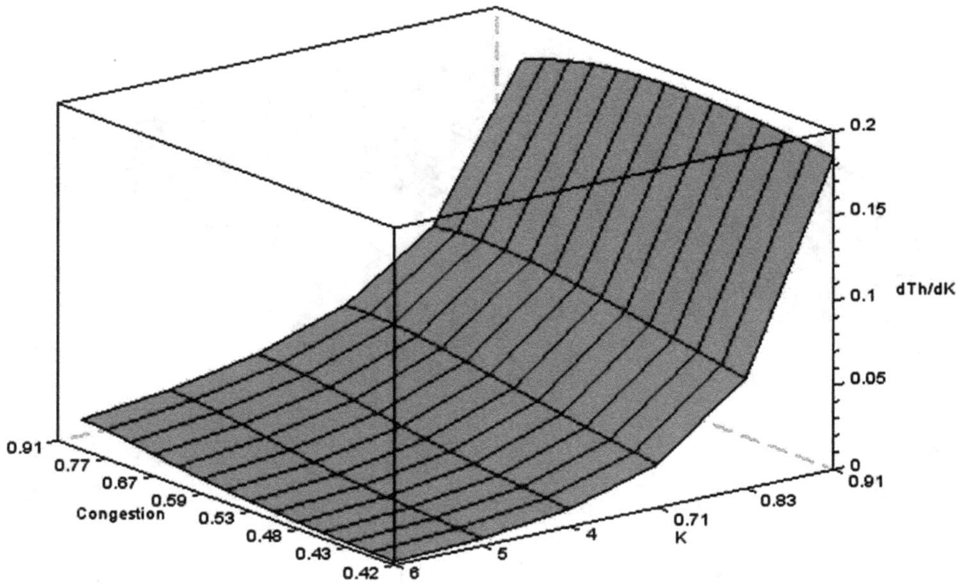

additional gain obtained decreases until a point eventually reached where allocating additional spaces is no longer justified as any additional benefit obtained would be marginal.

Figure 20 gives the results for a case where the decision variable is the congestion for some values of K. When the congestion increases (for example, because of increased demand) we can see that the additional benefit also increases. The additional benefit reaches a maximum value and then tends to drop. For the cases with K=3 and 4 the increment shows the same behavior, while for the particular case of K=1, as the congestion in the station increases, the possible additional benefit decreases.

General Aspects of the Buffer Allocation Problem

The Buffer Allocation Problem (BAP) is classified as a combinatory optimization problem and it is applied to the design of production lines. This problem has been studied for more than 50 years.

The proposed solutions implemented in the solution of the BAP involve the application of the combination of a generative method and an assessment method. The assessment method is used to get the value of the objective function

Figure 20. ΔTh/ΔK of a single M/G/1/K queue

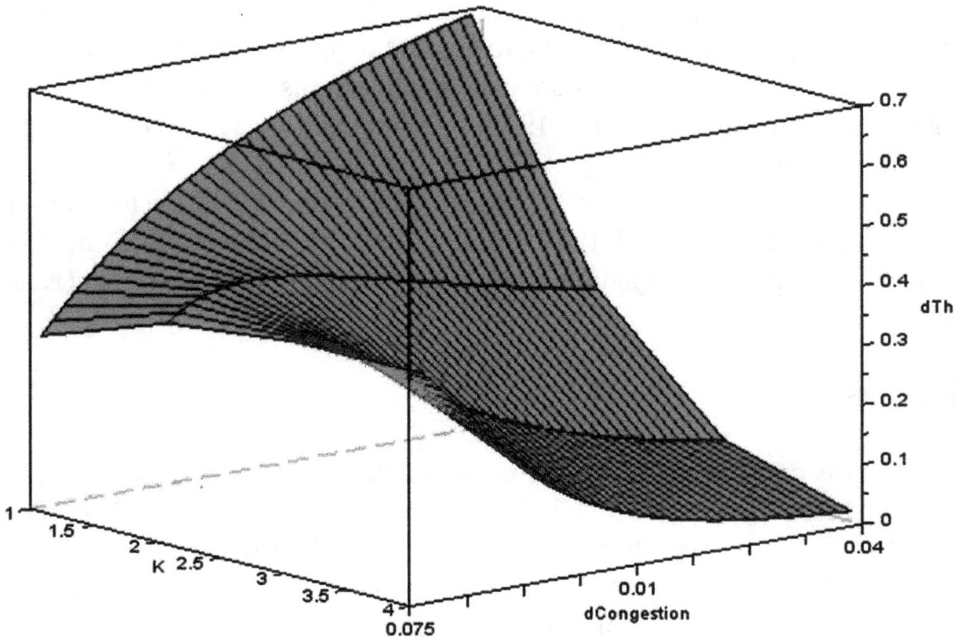

(*Th*, *WIP*, *TC*) from a set of inputs (*K* λ and μ), said value is communicated to the generative method whose purpose is to search for the optimum solution.

In recent years researchers have adopted several metaheuristic and search methods to find effective solutions for the BAP, which represents the essence of combinatory problems. Some methods that have been proposed for solving the BAP are:

1. Dynamic Programming (Yamashita and Altiok 1998).
2. Gradient (Gershwin and Schor, 2000 and Helber, 2001).
3. Conjugate directions (MacGregor Smith and Cruz, 2005)
4. Heuristics (Vouros and Papadopoulos, 1998; Fuxman, 1998; Harris and Powell, 1999; Papadopoulos and Vidalis, 2001; Hemachandra and Eedupuganti, 2003) and Aksoy and Gupta, 2010).

Metaheuristics Techniques

The metaheuristic techniques are a set of general algorithms for combinatory optimization whose main characteristic is that they are not designed for a specific problem, are general and very flexible procedures, which makes it possible to apply them to the vast majority of combinatory problems. Furthermore, these procedures make the search carried out by a heuristic method more efficient, allowing it to explore other regions of search space. The most common techniques are genetic algorithms (Michalewicz, 1992), tabu search (Glover and Laguna, 1993) and simulated annealing (Kirkpatrick, Gelatt and Vecchi, 1983).

Examples of metaheuristic techniques used to find a solution to the BAP are: Tabu Search (Lutz et al. 1998; Demir et al. 2012; Costa 2015), Simulated Annealing (Spinellis Papadopoulos 2000) and Genetic Algorithms (Kose 2010).

Exercises

1. Let us consider a station with two machines connected in series. The first station's service time is 3 minutes and the second station's service time is 3 minutes. There is space for 4 pieces between the two stations, so the total capacity of station two is 5 pieces. Get the cycle time, work in process and blocking probability of the system if the time between consecutive customers is 2.5 minutes. Let us assume that the stations are *M*/*M*/1 and *M*/*M*/1/*K* respectively.
2. For the above exercise, the cost of the work in process is \$3.5, while the cost of the service is \$4.0. Each piece produced is sold at \$50. Get the operating cost for the system.
3. Create a template on a spreadsheet for estimating the properties of a system in series with *M*/*M*/*c*/*K* stations. Use the method of Takahashi, et al (1980).
4. Get the cycle time and the amount of work in process for the next *M*/*M*/1/*K* stations.
   a. Arrival rate: 3 pieces an hour, service time: 0.4 hours with C^2_S = 2, *K*=3.
   b. Arrival rate: 2 pieces a minute, service time: 1 minute, C^2_S = 0.5, *K*=10.
   c. Arrival rate: 3 pieces a minute, service time: 2.75 minutes, C^2_S = 0.5, *K*=10.
   d. Arrival rate: 4 pieces a minute, service time: 1 minute, C^2_S= 0.5, *K*=10, waiting cost: \$ 5.0, service cost: \$5.0, revenues per piece sold: \$75.

e. For subparagraph d, estimate the value of K that maximizes the gain.
f. For subparagraph d, estimate the service time that maximizes the gain.

5. Create a template on a spreadsheet or modify the p_K program of the $M/G/2/K$ system for the $M/G/3/K$ case.

REFERENCES

Aksoy, H. K., & Gupta, S. M. (2011). Optimal management of remanufacturing systems with server vacations. *International Journal of Advanced Manufacturing Technology*, *54*(9-12), 1199–1218. doi:10.1007/s00170-010-3001-z

Akyildiz, I. F. (1988). Mean value analysis for blocking queueing networks. *IEEE Transactions on Software Engineering*, *14*(4), 418–428. doi:10.1109/32.4663

Altiok, T. (1997). *Performance analysis of manufacturing systems*. New York: Springer. doi:10.1007/978-1-4612-1924-8

Altiok, T., & Perros, H. G. (1987). Approximate analysis of arbitrary configurations of open queueing networks with blocking. *Annals of Operations Research*, *9*(6), 481–509. doi:10.1007/BF02054751

Balsamo, S., Nitto-Persone, V., & Onvural, R. (2001). *Analysis of Queueing Networks with Blocking*. Springer. doi:10.1007/978-1-4757-3345-7

Bolch, G., Greiner, S., de Meer, H., & Trivedi, K. S. (2006). *Queueing networks and Markov chains*. Wiley. doi:10.1002/0471791571

Buzacott, J. A., & Shanthikumar, J. G. (1993). *Stochastic models of manufacturing systems*. Upper Saddle River, NJ: Prentice Hall.

Castellucci, P. B., & Costa, A. M. (2015). A new look at the bowl phenomenon. *Pesquisa Operacional*, *35*(1), 75–72. doi:10.1590/0101-7438.2015.035.01.0057

Costa, A., Alfieri, A., Matta, A., & Fichera, S. (2015). A parallel tabu search for solving the primal buffer allocation problem in serial production systems. *Computers & Operations Research*, *64*, 97–112. doi:10.1016/j.cor.2015.05.013

Cruz, F. R. B., Duarte, A. R., & van Woensel, T. (2008). Buffer allocation in general single-server queueing networks. *Computers & Operations Research*, *35*(11), 3581–3598. doi:10.1016/j.cor.2007.03.004

da Silva, R. F., Sampaio, R. R., & Passos, F. U. (2016). Unbalancing Capacity: A Possible Way of Reducing Work-in-Process Inventory. *Journal of Industrial and Intelligent Information*, *4*(3), 198–203.

Demir, L., Tunalı, S., & Eliiyi, D. T. (2012). An adaptive tabu search approach for buffer allocation problem in unreliable non-homogenous production lines. *Computers & Operations Research*, *39*(7), 1477–1486. doi:10.1016/j.cor.2011.08.019

Fuxman, L. (1998). Optimal buffer allocation in asynchronous cyclic mixed-model assembly lines. *Production and Operations Management*, *7*(3), 294–311. doi:10.1111/j.1937-5956.1998.tb00458.x

Gelenbe, E. (1975). On approximate computer system models. *Journal of the Association for Computing Machinery*, *22*(2), 261–269. doi:10.1145/321879.321888

Gershwin, S. B. (1987). An efficient decomposition method for the approximate evaluation of tandem queues with finite storage space and blocking. *Operations Research*, *35*(2), 291–305. doi:10.1287/opre.35.2.291

Gershwin, S. B., & Schor, J. E. (2000). Efficient algorithms for buffer space allocation. *Annals of Operations Research*, *93*(1/4), 117–144. doi:10.1023/A:1018988226612

Glover, F., & Laguna, M. (1993). *Tabú search*. Springer.

Harris, J. H., & Powell, S. G. (1999). An algorithm for optimal buffer placement in reliable serial lines. *IIE Transactions*, *31*(4), 287–302. doi:10.1080/07408179908969834

Helber, S., Schimmelpfeng, K., Stolletz, R., & Lagershausen, S. (2011). Using linear programming to analyze and optimize stochastic flow lines. *Annals of Operations Research*, *182*(1), 193–211. doi:10.1007/s10479-010-0692-3

Hemachandra, N., & Eedupuganti, S. K. (2003). Performance analysis and buffer allocations in some open assembly systems. *Computers & Operations Research*, *30*(5), 695–704. doi:10.1016/S0305-0548(02)00034-5

Hillier, F. S., & Boling, R. W. (1966). The effect of some design factors on the efficiency of production lines with variable operations times. *Journal of Industrial Engineering*, *17*, 651–858.

Hillier, F. S., & So, K. C. (1991). The effect of the coefficient of variation of operation times on the allocation of storage space in production line systems. *IIE Transactions*, *23*(2), 198–206. doi:10.1080/07408179108963854

Hillier, F. S., & So, K. C. (1993). Some data for applying the bowl phenomenon to large production line systems. *International Journal of Production Research*, *31*(4), 811–822. doi:10.1080/00207549308956758

Hillier, F. S., & So, K. C. (1996). On the robustness of the bowl phenomenon. *European Journal of Operational Research*, *89*(3), 496–515. doi:10.1016/0377-2217(94)00287-8

Hillier, F. S., So, K. C., & Boling, R. W. (1993). Towards characterizing the optimal allocation of storage space in production line systems with variable processing times. *Management Science*, *39*(1), 126–133. doi:10.1287/mnsc.39.1.126

Hillier, M. (2013). Designing unpaced production lines to optimize throughput and work-in-process inventory. *IIE Transactions*, *5*(45), 516–527. doi:10.1080/0740817X.2012.706733

Hopp, W. J., & Spearman, M. L. (2008). *Factory physics*. Long Grove, IL: Waveland Press.

Kerbache, L., & MacGregor Smith, J. (1987). The generalized expansion method for open finite queueing networks. *European Journal of Operational Research*, *32*(3), 448–461. doi:10.1016/S0377-2217(87)80012-7

Kimura, T. (1996). A transform-free approximation for the finite capacity of *M/G/s* queue. *Operations Research*, *44*(6), 984–988. doi:10.1287/opre.44.6.984

Kirkpatrick, S., Gelatt, C. D. Jr, & Vecchi, M. P. (1983). Optimization by Simulated Annealing. *Science*, *220*(4598), 671–680. doi:10.1126/science.220.4598.671 PMID:17813860

Kose, S. Y., & Kilincci, O. (2015). Hybrid approach for buffer allocation in open serial production lines. *Computers & Operations Research*, *60*, 67–78. doi:10.1016/j.cor.2015.01.009

Lopez, C. E. (2014). *Unbalanced workload allocation in large assembly lines* (Master Thesis). Rochester Institute of Technology.

Lutz, C. M., Davis, K. R., & Sun, M. (1998). Determining buffer location and size in production lines using tabu search. *European Journal of Operational Research, 106*(2-3), 301–316. doi:10.1016/S0377-2217(97)00276-2

MacGregor Smith, J. (2003). M/G/c/K blocking probability models and system performance. *Performance Evaluation, 52*(4), 237–267. doi:10.1016/S0166-5316(02)00190-6

MacGregor Smith, J. (2004). Optimal design and performance modelling of M/G/1/K queueing systems. *Mathematical and Computer Modelling, 39*(9-10), 1049–1081. doi:10.1016/S0895-7177(04)90534-1

MacGregor Smith, J. (2011a). Properties and performance modelling of finite buffer M/G/1/K networks. *Computers & Operations Research, 34*(4), 740–754. doi:10.1016/j.cor.2010.08.014

MacGregor Smith, J. (2011b). The beauty of a good model. *Industrial Engineering (American Institute of Industrial Engineers), 43*(10), 34–38.

MacGregor Smith, J., & Cruz, F. R. B. (2005). The Buffer Allocation Problem for General Finite Buffer Queueing Networks. *IIE Transactions, 37*(4), 343–365. doi:10.1080/07408170590916986

Michalewicz, Z. (1992). *Genetic Algorithms + Data Structures = Evolution Programs*. Springer. doi:10.1007/978-3-662-02830-8

Papadopoulos, H. T., & Vidalis, M. I. (1998). Optimal buffer storage allocation in balanced reliable production lines. *International Transactions in Operational Research, 5*(4), 325–339. doi:10.1016/S0969-6016(98)00014-8

Perros, H. G., & Altiok, A. (1986). Approximate analysis of open networks of queues with blocking: Tandem configurations. *IEEE Transactions on Software Engineering, 12*(3), 450–461. doi:10.1109/TSE.1986.6312886

Shaaban, S., McNamara, T., & Hudson, S. (2014). Mean time imbalance effects on unreliable unpaced serial flow lines. *Journal of Manufacturing Systems, 33*(3), 357–365. doi:10.1016/j.jmsy.2014.02.006

Smut, T. L., & Perkins, W. C. (1985). Stochastic unpaced line design: Review and further experimental results. *Journal of Operations Management, 5*(3), 351–373. doi:10.1016/0272-6963(85)90019-1

Song, Y., & Takahashi, Y. (1990). Aggregate approximation for tandem queueing systems with production blocking. *Journal of the Operations Research Society of Japan*, *34*(3), 329–353. doi:10.15807/jorsj.34.329

Spinellis, D. D., & Papadopoulos, C. T. (2000). A simulated annealing approach for buffer allocation in reliable production lines. *Annals of Operations Research*, *93*(1/4), 373–384. doi:10.1023/A:1018984125703

Springer, M., & Makens, P. (1991). Queueing models for performance analysis and selection of single station models. *European Journal of Operational Research*, *58*(1), 123–145. doi:10.1016/0377-2217(92)90241-Z

Takahashi, Y., Miyahara, H., & Hasegawa, T. (1980). An approximation method for open restricted queueing networks. *Operations Research*, *28*(3), 591–602.

Timothy, L. U., & Wen-Chyuan, C. (2016). Designing energy-efficient serial production lines: The unpaced synchronous line-balancing problem. *European Journal of Operational Research*, *248*(3), 789–801. doi:10.1016/j.ejor.2015.07.015

Vouros, G. A., & Papadopoulos, H. T. (1998). Buffer allocation in unreliable production lines using a knowledge-based system. *Computers & Operations Research*, *25*(12), 1005–1067. doi:10.1016/S0305-0548(98)00034-3

Yamashita, H., & Altiok, T. (1998). Buffer capacity allocation for a desired throughput in production lines. *IIE Transactions*, *30*(10), 883–891. doi:10.1080/07408179808966542

Chapter 7
Special Topics in Queueing Theory

ABSTRACT

Topics covered in Chapter 7 are priority systems with preemptive or non-preemptive system, systems with N classes of customers, customers in groups: bulk arrivals, batch service, balking and reneging, and finite population. In a priority system, it is assumed that there are 1, 2, 3, …, N different classes or types of customers, where Type 1 customers are the most important while class N ones are the least important. When a server is available to serve a customer from the queue, the one with the highest priority level will go to the server to start their service process. In batch service, before starting the service process, a group or batch needs to be formed with a certain number of customers.

PRIORITY SYSTEMS

In the earlier chapters we studied systems assuming that the service priority is first in – first out, which, for want of a better expression, we could call "natural"; however we can often find systems where another service discipline is used, either because of the customers' requirements, the urgency of service, their level of importance or because the company has to achieve a particular objective.

Let's take the example of a hospital's A&E department. The customers arrive at the area in a random fashion and, once they are inside, ask to be

DOI: 10.4018/978-1-5225-5264-2.ch007

attended. Let us assume that at any given time someone with the symptoms of a cold arrives, immediately followed by someone whose life is at risk owing to a severe accident and who needs urgent attention.

If the FIFO discipline is respected, then the person with the cold would have to be served first and after they have been diagnosed and medicated then it would be the turn of the person involved in the accident. The point is that the life of the second patient depends on the speed with which he is given medical attention, whereas in the case of the first person, their symptoms are an indicator that they can wait without their life being put in danger. In this case, service priority is given to the person whose state of health is deteriorating at a faster rate. In a hospital, the care of the patient with a cold would be interrupted and service priority given to the person involved in an accident.

This example serves to illustrate a care system, known as "service priority" that is commonly used in hospital systems and, in general, in systems where the FIFO discipline is not suitable for use as a patient-care discipline.

In a priority system, it is assumed that there are 1, 2, 3, …, N different classes or types of customers, where type 1 customers are the most important while class N ones are the least important. When a server is available to serve a customer from the queue, the one with the highest priority level will go to the server to start their service process.

It is worth mentioning that when there are several customers of the same class, then the discipline that is applied to the subset of a particular class is FIFO. For the development of the respective model, we need to establish that the actions of the server, in the face of a new arrival, will be governed by one of the following considerations (Figures 1 and 2):

Figure 1. Preemptive. Customer of class 2 arrives and is attended but the service is interrupted when a client of class 1 arrives to the system.

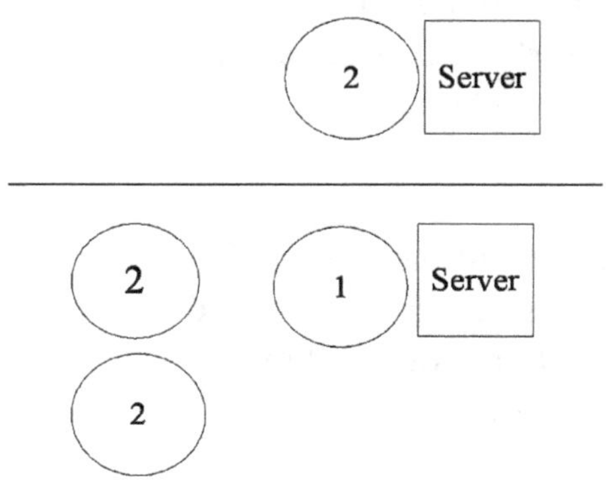

Figure 2. Non-preemptive. Customer of class 2 arrives and the service continues until the end. Customer of class 1 must wait until the end of service.

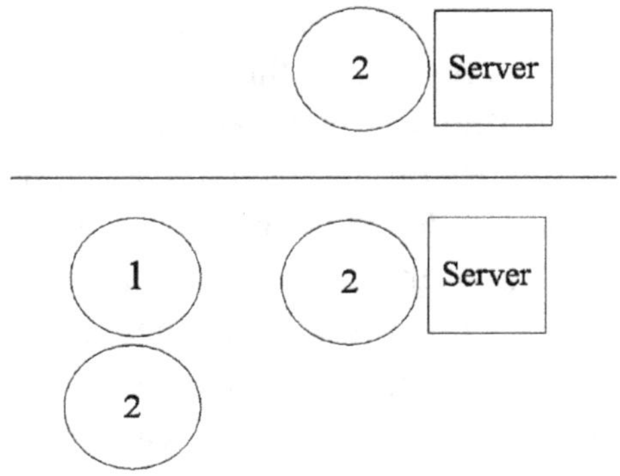

1. **Preemptive:** Once the customer enters the server, the service process starts. However if, during the process, a customer with higher priority or importance arrives then the care process is interrupted and the customer with the lowest priority is sent to the queue and the customer with higher priority enters.
2. **Non-Preemptive:** Once the customer has entered the server, the service process starts and is completed without interruptions.

M/M/1 System With Priorities

Let us assume that the customers arrive in the system and that the time follows an exponential probability distribution. Let us also assume that the flow of type 1 customers is λ_1 and the flow of class 2 customers is λ_1. In the end, the average service time 1/μ is similar for both types of customers.

Let the congestion for each type of customer be ρ_1 and ρ_2 respectively, where $\rho_1 = \lambda_1/\mu$ and $\rho_2 = \lambda_2/\mu$. Then in order for the system's queue not to grow uncontrollably, $\rho_1 + \rho_2 < 1$ has to be fulfilled.

Preemptive Service

Once a class 2 customer enters the server, the service process starts. However if a class 1 customer arrives during this period, the service process is interrupted and customer 2 is sent to the queue. For class 1 customers, the presence of class 2 customers is irrelevant.

The residence time (cycle time) in the system for a class 1 customer is given by the expression (Stewart, 2009):

$$CT_1 = \frac{1}{\mu\left(1-\rho_1\right)} \tag{1}$$

The amount of work in process is:

$$WIP_1 = \frac{\rho_1}{\left(1-\rho_1\right)} \tag{2}$$

When the service times are exponentially distributed then the number of customers in the system is independent of the order in which each class of customers have been served, therefore the total number of customers is the sum of the customers of each class, in other words:

$$WIP_T = WIP_1 + WIP_2 = \frac{\rho_1+\rho_2}{\left(1-\rho_1-\rho_2\right)} \tag{3}$$

Substituting the expression of the number of class 1 customers (2) in expression (3):

$$\frac{\rho_1}{\left(1-\rho_1\right)} + WIP_2 = \frac{\rho_1+\rho_2}{\left(1-\rho_1-\rho_2\right)} \tag{4}$$

Rearranging (4) to get the number of class 2 customers:

$$WIP_2 = \frac{\rho_1+\rho_2}{\left(1-\rho_1-\rho_2\right)} - \frac{\rho_1}{\left(1-\rho_1\right)} = \frac{\rho_2}{\left(1-\rho_1\right)\left(1-\rho_1-\rho_2\right)} \tag{5}$$

The cycle time of the customers of class 2 is obtained by applying Little's Law and (5):

$$CT_2 = \frac{\left[\dfrac{\rho_2}{(1-\rho_1)(1-\rho_1-\rho_2)}\right]}{\lambda_2} \quad (6)$$

Non-Preemptive Service

The class 1 customers have priority over the class 2 customers, however, once the care process starts for any of the two types of customers, it finishes without any interruptions.

The residence time of class 1 customers inside the system will be the sum of the time that the customer remains lined up (WIP_1/μ), plus the service time of customer 1, plus the fraction of time that the server attends to class 2 customers (ρ/μ), in other words (Stewart, 2009):

$$CT_1 = WIP_1\left(\frac{1}{\mu}\right) + \frac{1}{\mu} + \rho_2\frac{1}{\mu} \quad (7)$$

using Little's Law and expression (7) for the cycle time and the amount of work in process we get:

$$CT_1 = \frac{(1+\rho_2)/\mu}{1-\rho_1} \quad (8)$$

$$WIP_1 = \frac{(1+\rho_2)\rho_1}{1-\rho_1} \quad (9)$$

In the case of the class 2 customers, once they enter the server, the service continues until the requested service is completed, regardless of whether or not a customer with a higher priority has entered the system:

$$WIP_2 = \frac{(1-\rho_1(1-\rho_1-\rho_2))\rho_2}{(1-\rho_1)(1-\rho_1-\rho_2)} \quad (10)$$

For the cycle time:

$$CT_2 = \frac{\left(1-\rho_1\left(1-\rho_1-\rho_2\right)\right)\Big/\mu}{\left(1-\rho_1\right)\left(1-\rho_1-\rho_2\right)} \tag{11}$$

Example 7.1: In a factory, two types of pieces are produced (Table 1). The average service time follows an exponential probability distribution with a mean of 3 minutes. The type 1 piece arrives at the rate of 10 per hour and piece 2 has a rate of 7 per hour. The times between arrivals of each piece follow an exponential probability distribution.

Because of the characteristics of piece 2, the process can be interrupted to let type 1 pieces enter the process. Get the performance measures for both priority rules by applying expressions (8), (9), (10) and (11).

Observe that if the class 2 piece is allowed to complete its process, the average cycle time of the class 1 pieces in the system will be 35% longer. The average amount of work in process is 5.67 in both cases.

Systems With N Classes of Customers

The argument for getting an expression for the case with 1, 2, 3,…, *N* classes of customers is as follows: Let us assume that every class of customers arrives in the server at the rate of λ_i customers per unit of time. Now let us assume that the service time for every class of customer has a mean of $1/\mu_i$ and

Table 1. Performance of the system with both priorities

	1	2
	Preemptive	
Cycle time	0.1hrs.	0.67
WIP	1 piece	4.67 pieces
congestion	0.5	0.35
	Non Preemptive	
Cycle time	0.135	0.616
WIP	1.35	4.31

variance σ^2_i. Let the congestion for each type of customer be $\rho_i = \lambda_i/\mu_i$. Then, for the system to be considered stable, it must fulfill $\rho_1 + \rho_2 + \ldots + \rho_N < 1$.

Let the number of customers in the queue and the system be WIP_i and $WIP_{S,1}$ respectively. Furthermore, the queue cycle time will be denoted as $CT_{q,i}$ and the time left for the customer service process to end as $t_{res,i}$.

Case With Preemptive Service

While a class j customer is being served customers with equal or higher priorities arrive in the system, so the time required to serve all the customers will be made up by the following elements (Stewart, 2009):

1. The time taken to serve the customers with equal or higher priorities than customer j:

$$t_a = \sum_{i=1}^{j} \rho_i t_{res,i} + \sum_{i=1}^{j} \rho_i t_a \tag{12}$$

The first term is the time left to finish the service to the customer occupying the server, the second term is the sum of the service times of the customers in the same class that are in the system. Rearranging (12) to get t_a:

$$t_a = \frac{\sum_{i=1}^{j} \rho_i t_{res,i}}{1 - \sum_{i=1}^{j} \rho_i} \tag{13}$$

2. The time they will take to serve the customers with a higher priority than customer j:

$$t_b = \left(CT_{S,j} + t_{S,j}\right) \sum_{i=1}^{j-1} \rho_i \tag{14}$$

In equation (14), $CT_{q,j}$ is the time that customer j waits in the queue and $t_{S,j}$ is the service time for customer j. The cycle time that customer j stays in the queue is the sum of (13) and (14):

$$CT_{S,j} = t_a + t_b = \frac{\sum_{i=1}^{j} \rho_i t_{res,i}}{1 - \sum_{i=1}^{j} \rho_i} + \left(CT_{S,j} + t_{S,j}\right)\sum_{i=1}^{j-1} \rho_i \tag{15}$$

Finding $CT_{S,j}$ from equation (15):

$$CT_{S,j} = \frac{\sum_{i=1}^{j} \rho_i t_{res,i}}{\left(1 - \sum_{i=1}^{j} \rho_i\right)\left(1 - \sum_{i=1}^{j-1} \rho_i\right)} + \frac{\left(t_{S,j}\right)\sum_{i=1}^{j-1} \rho_i}{1 - \sum_{i=1}^{j-1} \rho_i} \tag{16}$$

If the inter-arrival time follows an exponential distribution, then $t_{res,i} = t_{Si}(C^2_S+1)/2$. If the service time follows an exponential probability distribution, then the station is M/M/1, so $C^2_S = 1$ and $t_{res,i} = t_{Si}$.

Case With Non-Preemptive Service

Once the priority *i* customer is being served, the process will continue to the end, therefore if a customer with a higher priority arrives, they will have to wait until the end of the process. Therefore the total residence time in the system (or CT_{Si}) will be the sum of the following periods (Stewart, 2009):

1. The time left to finish the service to the customer occupying the server, is also known as the remaining or residual time:

$$t_{res} = \sum_{i=1}^{N} \rho_i t_{res,i} \tag{17}$$

2. The sum of the service times of the customers in the same class that are in the system:

$$t_s = \sum_{i=1}^{N} WIP_{i,S} t_{s,i} \tag{18}$$

3. The sum of the service times of the customers with a higher priority that are in the system: For class 1 we have form (17) and (18):

$$CT_{S,i} = t_{res} + t_{serv} = \sum_{i=1}^{N} \rho_i t_{res,i} + WIP_{1,S} t_{S,1} \tag{19}$$

Applying Little's Law in expression (19) to obtain equation (20):

$$CT_{S,i} = \sum_{i=1}^{N} \rho_i t_{res,i} + \lambda_1 CT_{1,S} t_{S,1} = \sum_{i=1}^{N} \rho_i t_{res,i} + \rho_1 CT_{1,S} \tag{20}$$

Rearranging the expression to get the cycle time of the class 1 customers (21):

$$CT_{S,1} = \frac{\sum_{i=1}^{N} \rho_i t_{res,i}}{1 - \rho_1} \tag{21}$$

For class 2 and forward, it is necessary to calculate the system cycle time:

$$CT_{S,j} = \sum_{i=1}^{N} \rho_i t_{res,i} + \sum_{i=1}^{j} WIP_{i,S} t_{S,i} + CT_{S,j} \sum_{i=1}^{j-1} \rho_i \tag{22}$$

The third term in the equation (22) is the cycle time each class of customer takes in the system. Applying Little's Law:

$$CT_{S,j} = \sum_{i=1}^{N} \rho_i t_{res,i} + \sum_{i=1}^{j} CT_{S,i} \lambda_i t_{S,i} + CT_{S,j} \sum_{i=1}^{j-1} \rho_i = \sum_{i=1}^{N} \rho_i t_{res,i} + \sum_{i=1}^{j} CT_{S,i} \rho_i + CT_{S,i} \sum_{i=1}^{j-1} \rho_i \tag{23}$$

Rearranging (23):

$$CT_{S,j} \left(1 - \sum_{i=1}^{j-1} \rho_i \right) = \sum_{i=1}^{N} \rho_i t_{res,i} + \sum_{i=1}^{j-1} CT_{S,i} \rho_i + \rho_i CT_{S,j} \tag{24}$$

Observe that:

$$CT_{S,j} \left(1 - \sum_{i=1}^{j-1} \rho_i \right) = CT_{S,j-1} \left(1 - \sum_{i=1}^{j-2} \rho_i \right) \tag{25}$$

Then rearranging equation (25):

$$CT_{S,j}\left(1-\sum_{i=1}^{j-1}\rho_i\right)\left(1-\sum_{i=1}^{j}\rho_i\right)=CT_{S,j-1}\left(1-\sum_{i=1}^{j-2}\rho_i\right)\left(1-\sum_{i=1}^{j}\rho_i\right) \quad (26)$$

Rearranging formula (26) to calculate the cycle time:

$$CT_{S,j}=\frac{\sum_{i=1}^{N}\rho_i t_{res,i}}{\left(1-\sum_{i=1}^{j-1}\rho_i\right)\left(1-\sum_{i=1}^{j}\rho_i\right)};j=1,2,\ldots,N \quad (27)$$

When the service time follows an exponential probability distribution then the remaining time is equal to the service time $t_{S,i}$; therefore, for an M/M/1 priority system, expression (27) ends up as:

$$CT_{S,j}=\frac{\sum_{i=1}^{N}\rho_i t_{S,i}}{\left(1-\sum_{i=1}^{j-1}\rho_i\right)\left(1-\sum_{i=1}^{j}\rho_i\right)};j=1,2,\ldots,N \quad (28)$$

Example 7.2: A factory manufactures 4 types of pieces. Evaluate the queue cycle time of every product for each type of priority with expression (28). Let us assume that the station is M/M/1 (Table 2).

The calculations were carried out using the following code in Scilab:

Table 2. Cycle time and the amount of work in process for each class

Class	1	2	3	4
t_s (minutes)	1	1	1	1
λ (pieces/min)	0.25	0.25	0.25	0.15
CTq Preemptive	0.33 min.	1.66 min.	7 min.	39 min.
WIP preemptive	0.0825	0.415	1.75	9.75
CTq Non preemptive	1.2 min.	2.4 min	7.2 min	36 min.
WIP non preemptive	0.3	0.6	1.8	5.4

```
//preemptive
ro=lambda.*ts;:
sumro=0.0
sumro2=0.0
for a=1:N
 tcq(a)=0.0
end
for i=1:N
 j=1
 sumro=0.0
 while j<i
 sumro=sumro+ro(j)
 j=j+1
 end
 sumro2=0.0
 for jj=1:i
 sumro2=sumro2+ro(jj)
 end
 tcq(i)=sumro2*ts(i)/((1-sumro2)*(1-sumro))+ts(i)*sumro/
(1-sumro)
end
//non preemptive
sumro3=0.0
for i=1:N
 sumro3=sumro3+ro(i)*ts(i)
end
for i=1:N
 j=1
 sumro=0.0
 while j<i
 sumro=sumro+ro(j)
 j=j+1
 end
 sumro2=0.0
 for jj=1:i
 sumro2=sumro2+ro(jj)
 end
 tcq2(i)=sumro3/((1-sumro)*(1-sumro2))
end
```

Systems With C Parallel Servers

When the system is M/M/c, the expression for the cycle time of the class j customers with non-preemptive service is (Cobham, 1954; Wagner, 1969; Kella and Yechialli, 1985):

$$CT_{S,j} = \frac{ct_{S,i}}{\left(c - \sum_{i=1}^{j-1} \rho_i\right)\left(c - \sum_{i=1}^{j} \rho_i\right)} P_B; j = 1,2,\ldots,N \tag{29}$$

where P_B is the probability of the customer finding all the servers occupied (29), or in other words, the probability of the customer having to wait:

$$P_B = \frac{\rho^c}{c!\left(1 - \frac{\rho}{c}\right)} P_0$$

$$P_0 = \left[\sum_{k=0}^{c-1} \frac{\rho^k}{k!} + \frac{\rho^c}{c!\left(1 - \frac{\rho}{c}\right)}\right]^{-1} \tag{30}$$

In expression (30), P_0 is the probability of the system being empty. The following code in Scilab calculates the cycle time for the customers of class 1, 2,…, N in an M/M/c system.

```
//Non preemptive M/M/c
lambdat=sum(lambda)
ro=lambdat*ts/c;:
a=(lambdat*ts)^c;:
b=factorial(c)*(1-ro);:
suma=0.0
i=0
while i<c
 suma=suma+(lambdat*ts)^i/factorial(i)
 i=i+1
end
d=(lambdat*ts)^c;:
d=d/(factorial(c)*(1-ro));:
d=suma+d
p=(a/b)/d;:
for i=1:N
 j=1
 sumro=0.0
 while j<i
 sumro=sumro+lambda(j)*ts/c
 j=j+1
 end
```

```
 sumro2=0.0
 for jj=1:i
 sumro2=sumro2+lambda(jj)*ts/c
 end
 tcq(i)=(ts/c)/((1-sumro)*(1-sumro2))
 tcq(i)=p*tcq(i)
end
```

Example 7.3: Let us assume that a factory manufactures 4 types of pieces. The class 1 pieces have preference in the production, while class 4 ones have the lowest priority. The demand is for 15 pieces an hour. The factory has a work station with 3 machines in parallel, each machine processes pieces at the rate of 7 pieces an hour.

The proportion of pieces is as follows: class 1, 40%; classes 2, 3 and 4 20% each. So far we have been working under the rule that once the piece enters the machine, the process continues to the end without any interruptions.

By applying the above procedure we get the results in Table 3. We observe that the class 4 pieces are the ones stay longer waiting to be processed.

The Work Conservation Law (Kleinrock, 1965)

The service discipline is a control system that sets a series of characteristics or parameters for a customer that determine their position in a queue, however this position will depend on the arrival of a customer with a higher service priority.

When a customer arrives, they are allocated a priority p_1=1. This parameter will give the customer preference over any other with $p = 2, 3,\ldots$ and, in the event of a tie, in other words in the event that p_1=1 and p_2=1, then the order of care will be allocated, for example, using the first in – first out discipline.

Let us assume that:

1. All the customers that enter the system are served.

Table 3. Cycle time and the amount of work in process

Class	1	2	3	4
CTq	2.05 min	3.59 min	5.99 min	11.98 min
WIPq	0.205	0.179	0.299	0.599

2. There is a single server and this stays occupied as long as there are customers in the system.
3. If the service time follows an exponential distribution, then interrupting the service is admitted.
4. The inter-arrival time of customers follows an exponential probability distribution.

Let the total service demand (31) be the sum of the partial demands of each priority or class of customer:

$$\lambda = \sum_{i=1}^{P} \lambda_i \tag{31}$$

The service time for each priority is the sum (32) of the individual service times for the fraction of customers in each priority:

$$t_S = \sum_{i=1}^{P} \frac{t_{S,i}\lambda_i}{\lambda} \tag{32}$$

The total congestion of the system (33) will be the sum of the individual congestions:

$$\rho = \sum_{i=1}^{P} t_{S,i}\lambda_i = \sum_{i=1}^{P} \rho_i < 1 \tag{33}$$

Assuming that we are working with a priority system, the cycle time of each priority is CT_i; the product $\rho_i CT_i$ is the individual contribution of each priority and the sum of each contribution is:

$$\sum_{i=i}^{P} CT_{S,i}\rho_i = C \tag{34}$$

where for constant $C = \rho CT_{q,\ FIFO}$. The work conservation law (equation 34) establishes that the cycle time with priorities is equivalent to the weighted cycle time of a first in – first out system.

Example 7.4: Let us consider a factory with a machine and that works the first in – first out priority. The service demand is 10 pieces an hour. The machine works at a rhythm of 12 pieces an hour. There is a single server and the service time and interarrival time of customers follow an exponential probability distribution. The service discipline is first in – first out.

Let us assume that they decide to classify the pieces in 4 classes, where class 1 are pieces that have the highest preference and class 4 are the pieces with the lowest preference. The proportion for each piece is as follows: class 1, 60%; 15% each for class 2 and class 3 and 10% for class 4. It has been established that when the piece enters the machine it will be processed without any interruptions. The manager is interested in finding out what the average queue cycle time in the machine will be with this new working strategy.

The service time is the same for each piece.

The class 1 pieces spend, on average, less time waiting, while the class 4 pieces require about 1 hour 40 minutes queueing to be processed (Table 4).

Following on with Kleinrock's expression, each priority's average cycle time is multiplied by the each priority's congestion:

$$0.069 + 0.0462 + 0.0925 + 0.138 = 0.3472 \approx 20.832 \text{ minutes.}$$

When the system is operated using the first in – first out discipline, the cycle time is 0.4167 hours and multiplying it by the congestion we get

$$0.4167 * 0.8333 \approx 0.347 \approx 20.82 \text{ minutes.}$$

Example 7.5: Let us assume that a server is added in parallel. We shall again verify Kleinrock's conservation law for the non-preemptive operation.

Table 4. CT non-preemptive priority

Class	1	2	3	4
Queue cycle time (minutes)	0.138	0.37	0.74	1.66
$CT_q * \rho$	(0.138)(6/12) = 0.069	0.0462	0.0925	0.1388

Table 5. CT non-preemptive priority

Class	1	2	3	4
Queue cycle time (minutes)	0.0136	0.0198	0.023	0.028
$CT_q * \rho$	0.00134	0.00123	0.00148	0.0116

The sum gives us the result of 0.007294≈ 0.437 minutes. If the priorities are eliminated, then the queue cycle time is 0.0175 hours which, when multiplied by the congestion, gives us 0.0175 * 0.4167 ≈ 0.007291 (Table 5).

Priorities in Health-Care Systems ("Triage")

Outside the field of production and manufacturing, one of the most palpable examples of the application of a system that serves customers according to a level of priority is the health-care sector.

The word "triage" derives from the French word *trier*, which means to sort and referred to the classification and sorting of agricultural products. This word later was popularized in the field of health-care systems. The classification practice in health-care systems has its origin in the necessities of war. At the start of the 18th Century, military doctors developed a set of rules for allocating service priority to those injured in the battlefield; The formal application of this classification system is attributed to Baron Dominique-Jean Larrey, a doctor in Napoleon's army.

The rules are:

Those who are dangerously wounded should receive the first attention, without regard to rank or distinction. They who are injured in a less degree may wait until their brethren-in-arms, who are badly mutilated, have been operated and dressed, otherwise the latter would not survive many hours.

Later the naval doctor John Wilson proposed that doctors should serve the most gravely injured with the best possibilities of survival, leaving in the second place of priority those with minor or less severe injuries as well as the injured that would die regardless of whether or not they are given treatment.

At the present time the purpose of *Triage* is to provide quality care to people who have some type of medical emergency or need as well as to make sure that the workload is efficiently shared out. Nowadays the mostly commonly used classification systems are the Manchester Triage System (MTS),

Australian Triage Scale (ATS) and the Canadian Emergency Department Triage Scale(CEDTS) (Santos, Freitas and Martins, 2012).

For example, there are 5 categories in the Australian system's scale (Australasian College for Emergency Medicine, 2016): The highest priority (1) corresponds to patients in danger of losing their life and who require immediate medical attention, in other words, there is no waiting time. Category 2 is for patients with symptoms that indicate a need for urgent attention, such as exposed fractures or deep injuries. The patients in this category require a response time of 10 minutes. Category 3 is for patients with injuries or wounds that do not put their lives at risk and one characteristic these patients have is that they can walk by themselves. The recommended response time in this case is 30 minutes. Category 4 patients have symptoms such as breathing problems or bruises and the recommended waiting time is 60 minutes. Finally we have category 5, who are patients whose lives are not at risk and can wait for up to 120 minutes.

Every hospital faces the problem of demand exceeding its service capacity. In this regard the queueing theory approach is a tool that allows decision-makers to assess different proposals for speeding up the flow of patients in a system. In health-care systems, waiting is a key indicator in people's perception of the quality of the service.

Example 7.6: In a hospital the "Triage" follows the Australian Triage Scale. There are 3 doctors to attend to the patients in accordance with the classification allocated to them. The demand is 10 patients an hour and the time between patients follows an exponential probability distribution. According to the records, the proportion of patients is as follows: category 1, 5%; category 2, 10%; category 3, 50%; category 4, 20% and category 5, 15%.

The average service time spent on a patient is 20 minutes and follows an exponential probability distribution, once the patient sees a doctor, the attention is non-preemptive.

The director's perception is of a large number of patients arriving at the health unit and there are constant complaints about the service, particularly in respect of the waiting time. The number of patients with the highest priority is small and most of the demand is concentrated in categories 3 and 4.

The health unit's director needs an assessment of the average waiting time per type of emergency on the scale.

At the present time the health unit does not have enough capacity to attend to the flow of patients, as $\rho \approx 1.1$, therefore the demand cannot be met by the current number of doctors.

Table 6 shows the results of applying the procedure to estimating non-preemptive waiting times. A sensitivity analysis based on 4 doctors is presented in Table 6, estimating the waiting time in each case.

According to the results, at least 4 doctors are needed for the patients to flow inside the health unit (remember that adding parallel servers favors the flow inside the system). Comparing it with the ATS standard, urgency levels 2 – 5 would be within the standard; in respect of the highest level, adding 5 doctors, for example, would reduce waiting times by 1.35 minutes.

Example 7.7: There are other approaches for choosing the most suitable number of doctors. If the idea is to maintain the precise number of doctors to keep the system operating in a stable fashion, then we would recommend using 4 doctors.

Another strategy is to have the target of the probability of finding at least one doctor available being more than 0.5; then $1 - P_B > 0.5$. In this approach, 5 doctors would be needed.

Example 7.8: Let us assume that every doctor receives a salary of \$63 usd/ hour. The cost expression is

$$TotalCost = WIPC_{Wait} + \#ServersC_{Server};$$

Table 6. Estimated waiting time vs number of servers

Level of Urgency	$c = 4$	$c = 5$	ATS
1(highest)	3.43 min.	1.35 min	Immediately
2	3.92 min.	1.50 min.	10 min.
3	8.20 min.	2.56 min.	30 min.
4	24.60 min.	5.32 min.	60 min.
5 (lowest)	67.65 min.	9.04 min.	120 min
Probability of waiting (All servers busy)	0.65	0.32	
Congestion	0.833	0.667	

A customer being made to wait could result in loss of life, so the waiting cost term $WIPC_{Wait}$ is discarded in the analysis shown in Table 7.

Bulk Arrivals

In the models we have seen, it is assumed that the arrivals process consists of one customer at a time, and this is the same customer that is served once they are with the server. It has been observed that customers sometimes enter in groups (the readers themselves will have experienced this at some point) that join the queue and then when they are in front of the server, the attention is individual. In a university restaurant, students enter in groups of friends and join the queue, then each one is served separately when they reach the front of the queue (Figure 3).

The Kendall's notation for the Markov case is $M^{[x]}/M/c$, where [x] is the size of the group. Sometimes it is given an alternative batch name. The size of the group can be fixed or random.

Let us assume that the interarrival time $1/\lambda$ at a work station follows an exponential probability distribution. Likewise the service time $1/\mu$ is a random variable that is distributed exponentially. The service discipline is first in – first out. Let us assume that there is a single server.

The system of equations for the states of the system is:

Table 7. Total cost health unit

	$c = 4$	$c = 5$
Total Cost	\$ 252 usd	\$315 usd

Figure 3. Bulk arrival

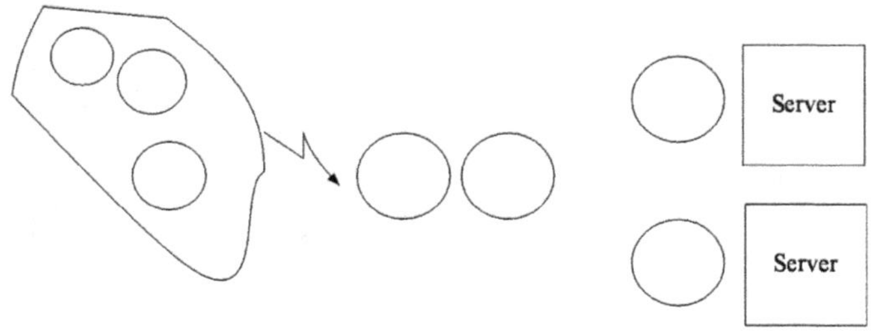

$$-\lambda p_0 + \mu p_1 = 0 \quad (35)$$

$$-\left(\lambda + \mu\right) p_n + \mu p_{n+1} = 0 \quad (36)$$

$$-\lambda p_{n-b} - \left(\lambda + \mu\right) p_n + \mu p_{n+1} = 0 \quad (37)$$

In order to resolve the above system (35, 36 and 37), a moment generating function is used (see, for example, White, Schmidt and Bennet, 1975). For the $M^{[x]}/M/1$ case, the expressions for the performance measures are set out below:

Congestion of the system:

$$\rho = \frac{\left[x\right]\lambda}{\mu} < 1 \quad (38)$$

The queue cycle time:

$$CT_S = \frac{1 + \left[x\right]}{2\mu\left(1 - \rho\right)} \quad (39)$$

The amount of work in process of the system:

$$WIP_S = \frac{\rho\left(1 + \left[x\right]\right)}{2\left(1 - \rho\right)} \quad (40)$$

Example 7.9: In a university restaurant there is a till where the customers' orders are received. The students usually arrive in pairs at the till and the person responsible for taking the order looks after one person at a time. The interarrival time is random with a mean of 2 minutes and follows an exponential probability distribution.

The cashier takes the order in 23.77 seconds and the response time is also a random variable that is exponentially distributed.

Get the performance measures of the system using formulas (38), (39) and (40).

Congestion is (2)(30)/151 .45= 0.396. System is stable. Cycle time is

(1+2)/(2(151.45)(1-0.396)) = 0.016 hours,

or about 59 seconds. WIP is

(0.396)(1+2)/((2(1-0.396)) = 0.98 clients.

Batch Service

Also industry often resorts to an operation in which, before starting the service process, a group or batch needs to be formed with a certain number of customers (Figure 4).

To start the processing of a batch, we select one of the following processing rules:

1. **Entire Batch Service Rule:** The service starts when the batch with β customers is complete.
2. **Partial Batch Service Rule:** Given a batch size with β customers, the processing starts when it reaches at least α customers ($\alpha < \beta$).

It will first be necessary to determine the average rate of batches that arrive in the system. The process starts with the formation of the batch that will enter the process. Said operation will require an amount of time to reach the specified size. Let us assume that the customers arrive at the station with an average interarrival time $t_a=1/\lambda$ and a coefficient of variation C_a^2. The first piece that arrives in the system has a time of $t_{a,1}$; in the case of piece 2, the time will be $t_{a,2}$; the time will be $t_{a,3}$ for piece 3 and so on for the rest of the pieces that make up the batch. The total time required to make up the batch will be the sum of the times between arrivals (Curry and Feldman, 2011):

$$t_a = t_{a,1} + t_{a,2} + t_{a,3} + \ldots + t_{a,\beta} \quad (41)$$

If the times between arrivals are independent random variables then the average time required to make up the batch will be:

Figure 4. Customers arrival to form the batch before the operation in the station

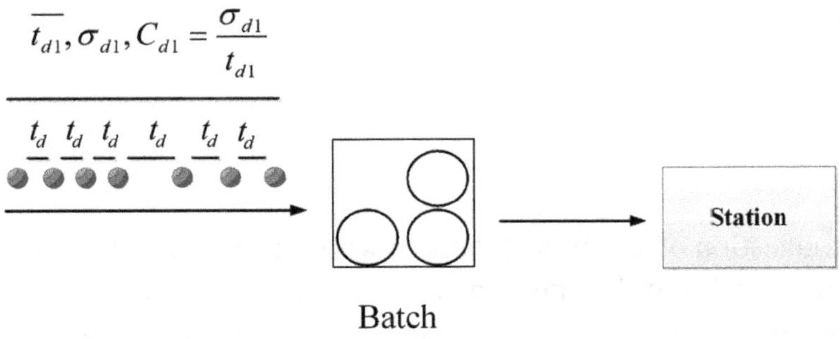

$$t_\beta = \beta t_a \tag{42}$$

Combining the expressions (41, 42) to get the batch arrival rate we get:

$$\lambda_\beta = \frac{1}{t_\beta} = \frac{1}{\beta t_a} = \frac{\lambda}{\beta} \tag{43}$$

Full Batch Service Rule

In this first part, we shall assume the entire bulk service rule to develop the respective analytical model. Once the batch has β pieces then the server starts the service process. The squared coefficient of variation of the batch interarrival times is:

$$C^2_{a,b} = \frac{C^2_a}{\beta} \tag{44}$$

To process each batch we need time to set up the station, $t_{set\text{-}up}$, each piece requires a processing time with a mean of t units of time and a coefficient of variation C^2_S. The total time for processing a full batch is the sum of the processing times of β pieces that make it up plus preparation time, in other words:

$$t_\beta = \beta t_S + t_{set-up} \tag{45}$$

The squared coefficient of variation corresponding to the processing of a full batch is:

$$C_{S,\beta}^2 = \frac{C_S^2}{\beta} \tag{46}$$

The congestion of a station that processes batches of pieces is $\rho_\beta = \lambda t_\beta$. Let λ / β be the number of batches per unit of time, then a system's congestion by batches is $\rho_b = (\lambda / \beta)\ t_\beta$. Substituting the time expression to process a batch(45):

$$\rho_b = \frac{\lambda}{\beta}\left(\beta t_S + t_{set-up}\right) \tag{47}$$

Reorganizing:

$$\rho_b = \lambda\left(t_S + \frac{t_{set-up}}{\beta}\right) < 1 \tag{48}$$

Unlike the systems seen in the above chapter, observe that (47) and (48) take into account the time for the preparation of a batch. Moreover, this time is often a source of variability in a system and can seriously affect the performance measures of a work station.

System Cycle Time

Before calculating each batch's cycle time in the system, we must consider what happens to it on leaving the station:

1. The batch joins up again before being sent to another station.
2. The batch is dissolved and each piece is treated as a new customer.

For the first case, the cycle time is the sum of the batch queue cycle time, plus the time that each piece has to wait for its turn to be processed in the station plus the time that each piece has to wait for the batch to join together again (Curry and Feldman, 2011):

$$CT_b = CT_{q,b} + t_{Wait} + t_{service} + t_{batching} \tag{49}$$

In the case of the time each piece has to waits to be processed, the reasoning is as follows: the first piece that enters the process does not have to wait its turn, therefore t_{wait},1= 0; the second piece has to wait the time it takes the first to be processed, $t_{s,2}$, in other words $t_{s,2} = t_{s,1}$ and the third piece must wait $t_{s,1} + t_{s,2}$; then (Curry and Feldman, 2011):

$$t_{Wait} = t_{S,1} + \left(t_{S,1} + t_{S,2}\right) + \left(t_{S,1} + t_{S,2} + t_{S,3}\right) + \ldots + \left(t_{S,1} + t_{S,2} + t_{S,3} + \ldots + t_{S,\beta-1}\right) \tag{50}$$

The service times are independent random variables. The above expression is rearranged as follows:

$$t_{Wait} = \left[\left(\beta - 1\right) + \left(\beta - 2\right) + \ldots + 1\right] t_S \tag{51}$$

$$t_{Wait} = \frac{\left(\beta - 1\right) t_S}{2} \tag{52}$$

If the batch has to be put together again on leaving, then the time spent waiting for it to be put together has the same expression (52). Finally, the queue cycle time will be calculated using Kingman's approximation of a G/G/1 system:

$$CT_q \approx \left(\frac{C_{a,\beta}^2 + C_{S,\beta}^2}{2}\right)\left(\frac{\rho_\beta}{1 - \rho_\beta}\right)\left(\beta t_S + t_{set-up}\right) \tag{53}$$

The total batch cycle time, substituting the terms corresponding to the waiting time, the batch forming time and the service time in (53) (Curry and Feldman, 2011):

$$CT_{S,b} \approx \left(\frac{C_{a,\beta}^2 + C_{S,\beta}^2}{2}\right)\left(\frac{\rho_\beta}{1 - \rho_\varphi}\right)\left(\beta t_S + t_{set-up}\right) + \frac{\left(\beta - 1\right) t_S}{2} + \left(\beta t_S + t_{set-up}\right) + \frac{\left(\beta - 1\right) t_S}{2} \tag{54}$$

Observe in expression (54) that when the batch size is β=1, it insurance same as when the customers are served immediately and the terms corresponding to the waiting in the queue and taking turn in the process are zero; in this same situation, the quadratic coefficients of variation will be $C_{a,\beta}^{\ 2} = C_{a,}^{\ 2}$ for the arrivals and $C_{S,\beta}^{\ 2} = C_S^2$ for the service time.

Example 7.10: Let us consider a factory where the operation is carried out using batches of 10 pieces. All the pieces arrive from another stage in the process with an average spacing of 0.15 hours and variability $C_{a,}^{\ 2}$ = 2.5. The preparation of each batch requires 0.1 hours and as soon as it is ready, the operation starts in the machine. The processing time for each piece requires 6 minutes with a variability C_S^2= 1.

Get the system cycle time. The batch queue cycle time:

$$CT_q \approx \left(\frac{0.25+0.1}{2}\right)\left(\frac{0.73}{1-0.73}\right)\left(10\times 0.1+0.1\right)=0.53hrs \tag{55}$$

The waiting time and the batch forming time are respectively:

$$t_{Wait}=t_{Batching}=\frac{\left(10-1\right)\times 0.1}{2}=0.45hrs \tag{56}$$

Substituting in expression:

$$CT_{S,b}\approx\left(0.53\right)\left(10\times 0.1+0.1\right)+0.45+\left(10\times 0.1+0.1\right)+0.45\approx 2.531hrs. \tag{57}$$

If the batch no longer has to form a queue on leaving the station then only the time waiting to be processed needs to be taken into account, therefore the cycle time will be 2.08 hours.

Results in (55), (56) and (57) were obtained applying expressions (52), (53, (54).

Example 7.11: The expression for the system cycle time is hard to derive for the batch size that minimizes the system cycle time. It is faster instead to apply a numeric method without assessing derivatives or plotting the function for values of β (Table 8).

Table 8. Cycle time vs. batch size

β	ρ	$CT_{S,\beta}$
3	0.89	2.47
4	0.83	1.89
5	0.80	1.84
6	0.78	1.91
7	0.76	2.04
8	0.75	2.19
9	0.74	2.36
10	0.73	2.53
11	0.73	2.71
12	0.72	2.89
13	0.72	3.08
14	0.71	3.27
15	0.71	3.46
16	0.71	3.65
17	0.71	3.84
18	0.70	4.04
19	0.70	4.23

As can be appreciated in Figure 5, under the station's operating conditions, the system cycle time can be lowered from 2.53 to 1.91 hours using batches of 6 pieces.

When the system has exponentially distributed interarrival times and service time and the system is $M/M^{[x]}/1$ in Kendall's notation, then the expression is:

$$CT_{S,b} \approx \left(\frac{1}{\beta}\right)\left(\frac{\rho_\beta}{1-\rho_\beta}\right)\left(\beta t_S + t_{set-up}\right) + \frac{(\beta-1)t_S}{2} + \left(\beta t_S + t_{set-up}\right) + \frac{(\beta-1)t_S}{2} \tag{58}$$

With the data in example 11 and by applying the expression (58), the optimum batch size is 4 pieces and the cycle time will be 1.43 hours.

One aspect to be highlighted is that the queue cycle time gets shorter as the batch size increases, Whereas, for example, while the process time for the complete batch begins to dominate the time every piece spends waiting

Figure 5. Cycle time vs. lot size

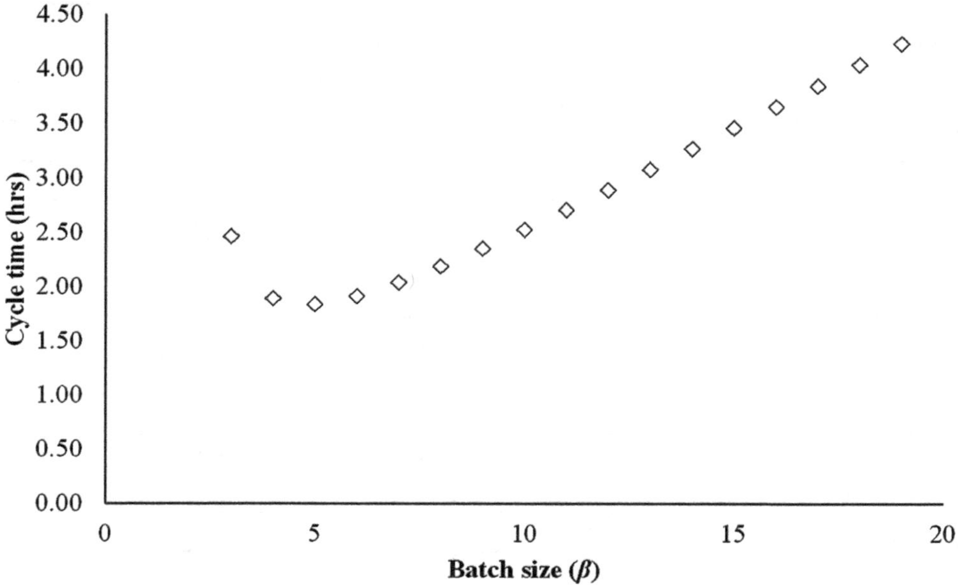

to enter the process and the time spent waiting to put the batch together again as it leaves.

For the case where there are a several parallel servers, expression (53) can be modified as follows:

$$CT_{S,\beta} \approx \left(\frac{C_{a,\beta}^2 + C_{S,\beta}^2}{2}\right)\left(\frac{\rho_\beta^{\sqrt{2c+2}-1}}{c\left(1-\rho_\beta\right)}\right)\left(\beta t_S + t_{set-up}\right) + \frac{\left(\beta-1\right)t_S}{2} + \left(\beta t_S + t_{set-up}\right) + \frac{\left(\beta-1\right)t_S}{2} \tag{59}$$

Once again, in the case of β=1, expression (59) corresponds to a G/G/c system that we explained earlier.

To close this section, the expressions reported in Bolch, Greiner, de Meer and Trivedi (2006) are given below. In this text, the authors apply the approximations of Allen – Cunneen and Kramer- Langerbach – Belz to the development of a couple of formulas for estimating the average queue size in a $G/G^{[x]}/m$ system.

Expression with the Allen – Cunneen approximation:

$$WIP_q = \left(\frac{C_{a,\beta}^2 + C_{S,\beta}^2}{2}\right)\frac{\rho_\beta p_m}{1-\rho_\beta} \tag{60}$$

In formula (60), p_m is the probability of the system being occupied:

$$p_m = \begin{cases} \frac{\rho_\beta{}^m + \rho_\beta}{2} & \rho_\beta > 0.7 \\ \rho_\beta^{\frac{m+1}{2}} & \rho_\beta < 7 \end{cases} \tag{61}$$

Expression (61) is also an approximation. Expression with the Kramer-Langerbach – Belz approximation:

$$WIP_q = \left(\frac{C_{a,\beta}^2 + C_{S,\beta}^2}{2}\right)G \tag{62}$$

The term G in formula (62) is:

$$G = \begin{cases} \exp\left(-\frac{2}{3}\left(\frac{1-\rho}{p_m}\right)\left(\frac{\left(1-C_a^2\right)^2}{C_a^2 + C_S^2}\right)\right) & 0 < C_a^2 \le 1 \\ \exp\left(-\left(1-\rho\right)\frac{\left(C_a^2 - 1\right)}{C_a^2 + 4C_S^2}\right) & C_a^2 > 1 \end{cases} \tag{63}$$

Observe in (63) the influence of the natural variability of the arrivals and the service time. The cycle time is calculated using approximation (64):

$$CT_{Batch} \approx \beta WIP_q + \frac{\beta - 1}{2} \tag{64}$$

Sensitivity Analysis

The improvement in an operation's cycle time should take several factors into account at the same time. As we have already seen in the earlier chapters, as well as taking the service time or the batch size into account, we need to take the variability of the system as a factor for consideration.

For example look at Figure 6 where we can see the behavior of the station's cycle time in respect of batch size β=(3 – 20) and the variability of arrivals C^2_a = (2.5, 2, 1.5, 0.5). The optimum batch size for the original data is $\beta_{opt,1}$=6 pieces with a cycle time of 1.91 hours. Let us assume that the variability of arrivals is decreased to C^2_a= 1.0, the new result is $\beta_{opt,2}$=4 pieces and the cycle time is 1.43 hours.

Figure 6 also reveals a quasi-convex function of the cycle time in respect of the batch size, it is also worth waiting for an improvement in the cycle time in respect of the arrival variability in this case.

Another variable of interest is the reduction in preparation time. Figure 7, using the data from example 11, shows the effect. We observe that when the preparation time is high, the system becomes unstable, in other words the demand is higher than the service capacity and, in consequence, the queue grows boundlessly. The figure corresponds to the truncated section (the results of the expression are meaningless in this phase). For every preparation time value there is a different optimum batch size, using the example's data we find that with every increment of 0.05 hours in the preparation time, the optimum batch increases by a unit.

Figure 6. Cycle time vs. variability arrivals, lot size (β)

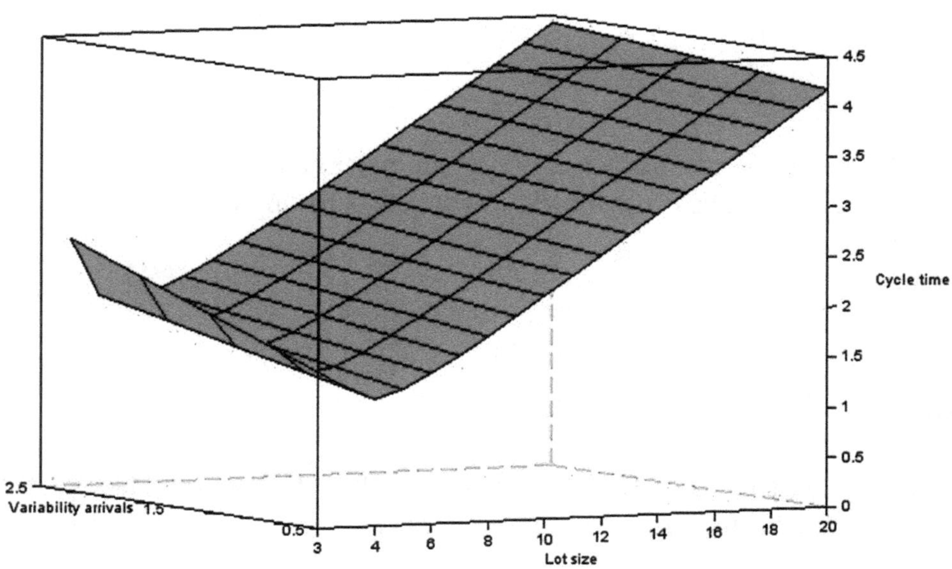

Figure 7. Cycle time vs. lot size, set up time

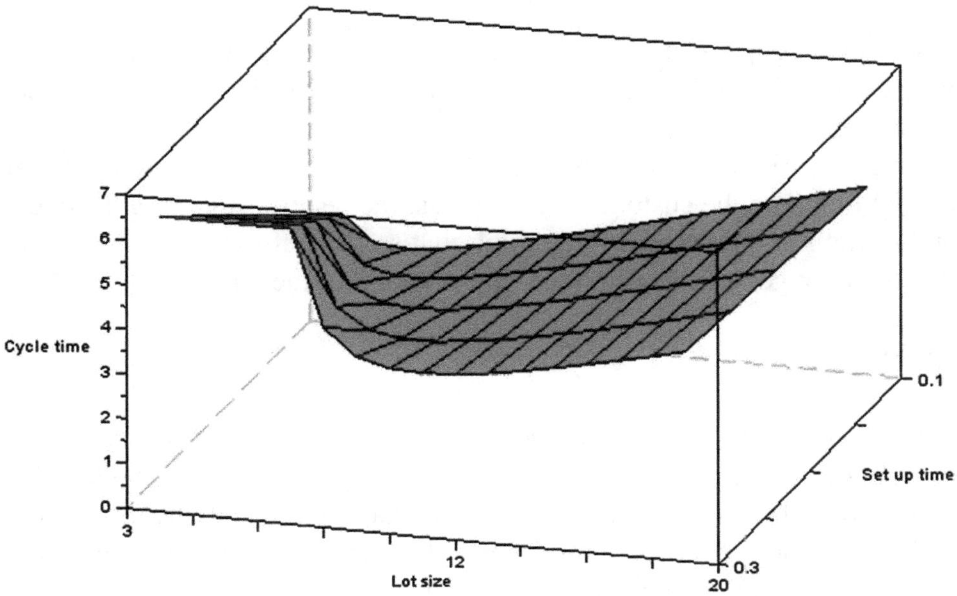

Figure 8 shows the rate of change in the cycle time in respect of the change in the preparation time. As changes are made in the batch size, the cost drops. The change will be gradually decreasing as the batch size approaches the optimum result.

Figure 8. ΔCT/Δset-up time vs. set up time

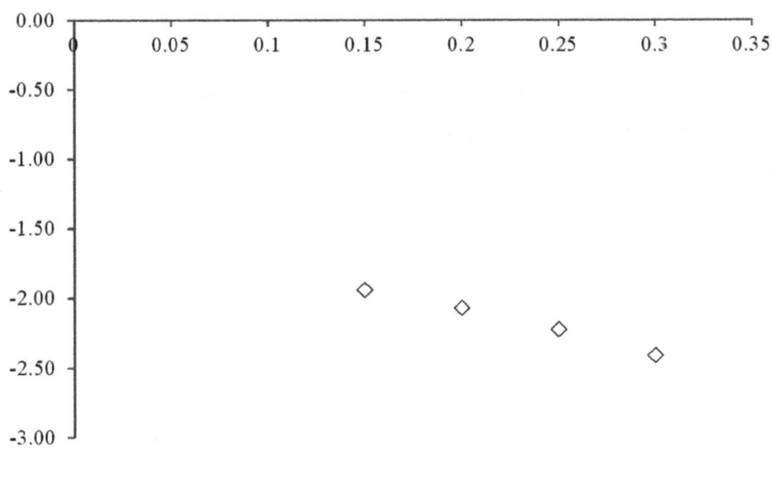

Balking and Reneging

There are a couple of phenomena that are typical of queueing systems and very evident when directly working with people: balking or joining queue and reneging.

The first arises during the process of the customers arriving in the system. Before lining up the customer will evaluate the benefit of joining the queue and waiting to be served, if they think that said benefit is minimum or that they will not get any benefit, then they do not join the queue and leave.

Balking

Let us consider a system where the customers arrive at a rate of λ customers per unit of time and the interarrival times are exponentially distributed. The service time $1/\mu$ is also a random variable that is exponentially distributed.

Let δ be the proportion of customers that decide to join the queue. Then the rate of customers that enter the system and line up is $\delta \cdot \lambda$, the system's congestion will be $\rho=\delta\ \lambda/\mu < 1$. The probability of finding n customers in the system will be:

$$p_n = \left(1-\rho\right)\rho^n \tag{65}$$

Now let us assume that the number of customers that decide to line up is inversely proportional to the number of customers present in the system. Let the probability of finding the system empty be:

$$p_0 = \exp\left(-\rho\right) \tag{66}$$

where $\rho=\lambda/\mu$. Combining (65) and (66) to determine the distribution of the number of customers in the system n will be:

$$p_n = \frac{1}{n!}\left(\frac{\lambda}{\mu}\right)^n p_0 \tag{67}$$

or else:

$$p_n = \frac{\rho^n}{n!} \times \exp(-\rho) \tag{68}$$

Example 7.12: Let us assume that the students arrive at the till of a cafeteria at the rate of 15 customers an hour. There is just one person and the time taken up in taking the order, charging and giving back change is 3 minutes. Let us assume that the proportion of customers during that period that line up in the queue for the till is 90%. Estimate the distribution of the number of customers n in the system without taking the balking into account, assuming δ = 0.90 and assuming that the rate of arrival in the system is inversely proportional to the number of customers in the system.

Figure 9 compares the probabilities for the case without considering balking with the case where the probability is assumed to be inversely proportional to the number of customers in the system given by expressions (66), (67) and (68).

Figure 10 shows the effect of balking on the three cases. Although it is obvious that the probability of finding (for example 10 customers in the system) is lower if we consider that a fraction of the customers decides not to line up, we need to remember that they represent a revenue or sale that was

Figure 9. Probability distribution of n customers in the system

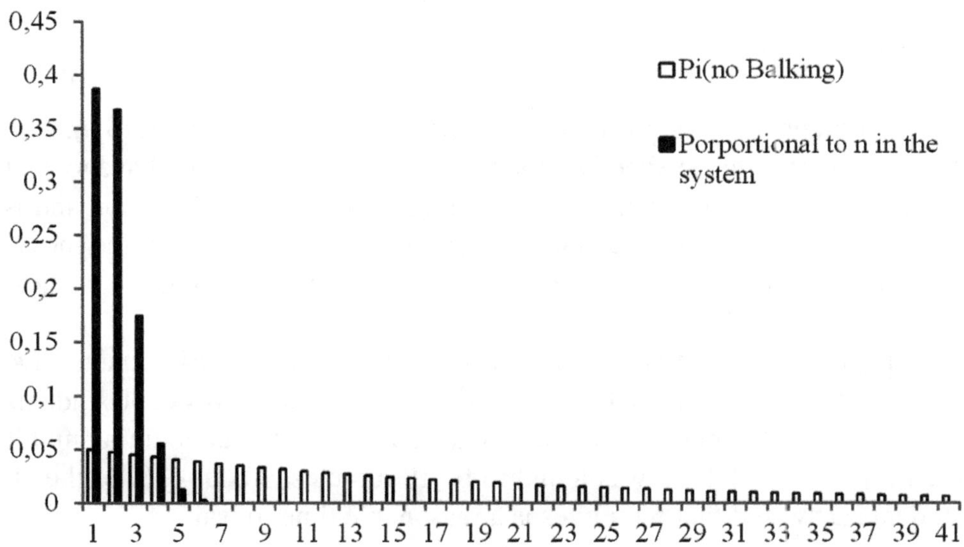

Figure 10. Probability of finding n customers in the system, high congestion (ρ=0.95)

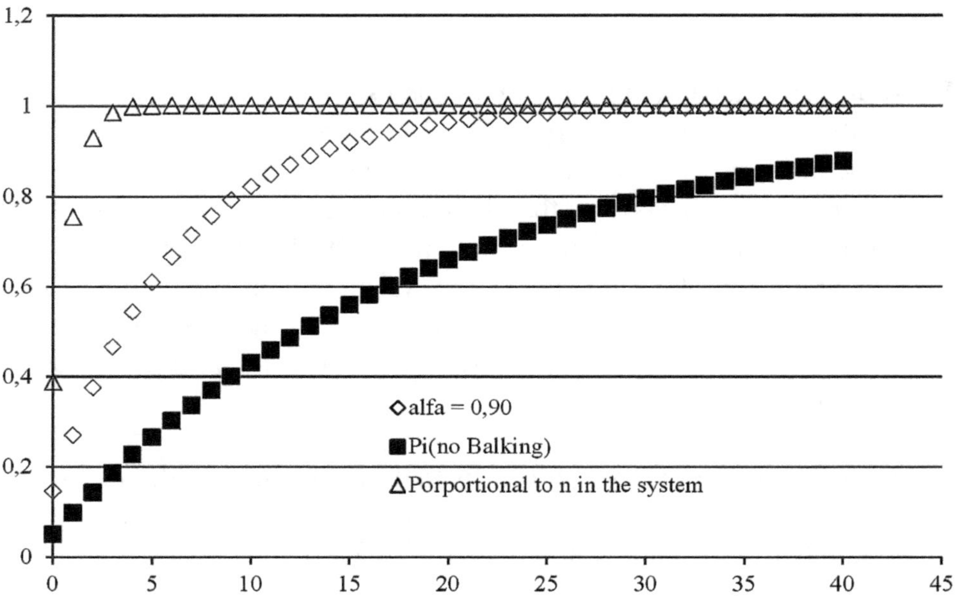

not obtained, hence the applicability of expressions (65), (66), (67) and (68). When expression (68) is applied, we observe that the probability of being a large number of customers is low in comparison to the results obtained, assuming that all the customers decide to enter.

If the congestion is low (for example because of a drop in demand) then the three expressions show the behavior of Figure 11.

Reneging

The second phenomenon is reneging. Once the customer decides to line up, they do so from the perspective, for example, that they will not have to wait too long. The level of tolerance for waiting depends on each person and is related to the type of service they expect to receive: queueing at the movie theater to see a much acclaimed movie is not the same as waiting in a medical clinic or a hospital.

It is possible to get an idea about this level of tolerance in quantitative terms by means of a survey. For example, in an informal survey of 30 students at a university, the maximum time that a student could tolerate lining up for the till of their cafeteria was found to be 10 minutes. About 43% of the 30 students answered that 5 minutes is a reasonable time to wait.

Figure 11. Probability of finding n customers in the system, low congestion (ρ=0.5)

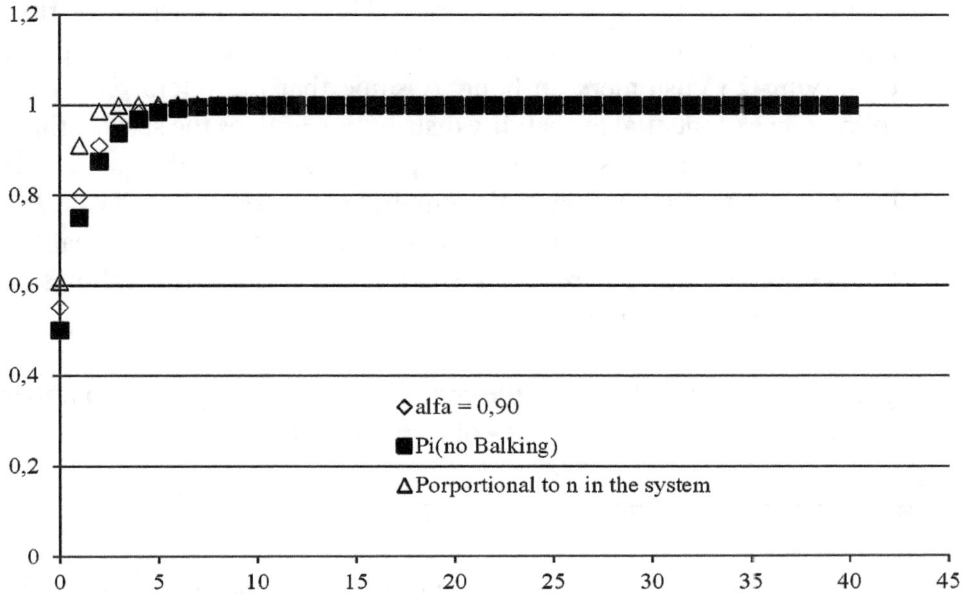

It is of the utmost importance to quantify the probability of a customer reneging or leaving the queue in systems such as call-centers, as these events imply services that have not been provided. In order to manage this class of systems, it is important to find out the probability distribution of the number of customers in the system, the probability of a customer reneging before being served and the evolution of the system over time (Ward, 2011).

Let us assume that the customers arrive at a rate of λ customers per unit of time and the interarrival times are exponentially distributed. The service time $1/\mu$ is also a random variable that is exponentially distributed. Assume that the customers leave the system at a rate of ξ customers per unit of time and this is independent of the arrival rate and service rate.

The speed at which customers leave the system will be the sum of the rate at which customers are served plus the reneging rate $\mu + \xi$. The system's congestion will be $\rho = \lambda / (\mu + \xi)$.

The probability distribution for the number of customers in the system n will be:

$$p_n = (1-\rho)\rho^n \tag{69}$$

Example 7.13: Let us assume that there is one till in a cafeteria where the customers places their orders for food. The demand during the rush hour is 20 customers an hour and there is just one person who is capable of serving 25 customers an hour. Assume that the interarrival time follows an exponential probability distribution as does the service time. Assume that the some students get impatient and leaves the queue. Get the probability distribution of the number of customers in the system with equation (69) for different values of $1/\xi$. Let us assume that the time between customers reneging is a random variable that follows an exponential probability distribution.

Figure 12 shows the behavior of the cumulative probability for the number of customers in the system for values of ξ =(0,1,3,6).

Final Comment on Priority Systems

There is an abundance of literature about priority systems, batches as well as about with systems with balking and reneging. In the case of priority systems, the models are often being applied to support decision-making aimed at

Figure 12. Cumulative probability of reneging vs rate of reneging, n customers

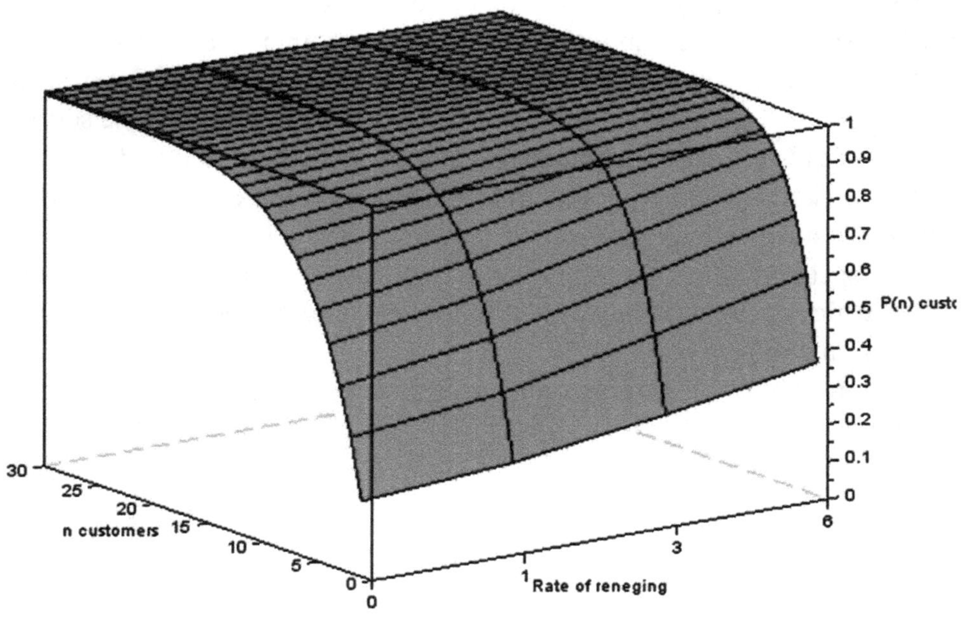

improving the perception of the service in hospital systems, improving the flow of patients and dealing with the growing demand for care (Hernández-González, et al, 2016).

With respect to batches systems, la expression given in this chapter corresponds to a single station. In Curry and Feldman (2011) a complete chapter is devoted to getting models for series and network systems. Hopp and Spearman (2008) consider a model for systems with batches that have some simplifications. The text of Bolch, Greiner, de Meer and Trivedi, (2006) is another source of methods for this problem.

Given that in a system with several work stations it has to work with a considerable number of variables and constraints, it is a good idea to work under an optimization paradigm for finding the optimum batch size for the system and to use a procedure for finding the solution. One analytical strategy would be to apply the concept of economic bottleneck proposed by Lawrence and Buss (1995).

The balking and reneging systems have been studied in the area of call-centers, as one performance measurement is precisely the number of customers that do not enter a system and the number of customers that leave the queue out of impatience. Papers by Chauhan and Sharma (1996), Ward and Glyn, (2005), Whitt (2006), Atar, Giar and Shimkin (2010), Dai and He (2010) are some recent examples, models and theoretical aspects of these types of systems.

Finite Population

A queueing system with finite population (or source-finite) model is a system in which there are a limited number of M customers who use the service offered by the system. In this model, arrival speeds change according to the number of clients (n) in the system $\lambda_n = (M - n)\,\lambda$. Since $\lambda_n = 0$ when $n = M$ this model will eventually reach a steady-state condition.

The usual application of this model has been to the machine repair problem, where one or more maintenance people are assigned the responsibility of maintaining in operational order a certain group of M machines by repairing each one that breaks down. For derivation of equations for the finite population model see (Cooper, 1981, pp. 111-116)

To analyze the performance of a (M/M/c)(GD/K/m) queue we can make use of "queueing package" in the R statistical programming language. The command is

```
NewInput.MMCKM(lambda=0, mu=0, c=1, k=1, m=1, method=0)
```

The arguments are: *lambda*: arrival rate, *mu*: server service rate, *c*: number of servers (repairmen), *k*: system capacity, *m*: population size (Machines). (*m* >= *K*).

- **Method:** Method of computation of the probabilities of *K* (system capacity) customers down. With method=0, the exact results are calculated using the formal definition. With method=1, approximate results are calculated using Stirling approximation of factorials and logarithms.

Example 7.14: One crew is responsible for 10 machines. Machines run on the average for 20 minutes then require an average 5 minutes service period, both times exponentially distributed with λ = 3 units/hour and μ = 12 units/hour.

Obtain a comparison of the performance measures for different number of workers in the crew (2, 3, 4).

To analyze the performance of M/M/c/K/m queue we then do the following:

```
x2<-NewInput.MMCKM(lambda=3,mu=12,c=2,k=10,m=10,method=0)
x3<-NewInput.MMCKM(lambda=3,mu=12,c=3,k=10,m=10,method=0)
x4<-NewInput.MMCKM(lambda=3,mu=12,c=4,k=10,m=10,method=0)
y2=QueueingModel(x2)
y3=QueueingModel(x3)
y4=QueueingModel(x4)
CompareQueueingModels(y2, y3,y4)
```

The NewInput.MMCKM function specifies the input parameters for the M/M/c/K/m queue, while the QueueingModel(x) function builds the queueing model. CompareQueueingModels(y2,y3,y4) gives comparative analysis of the queue.

Table 9. Output of R program

Model	λ	μ	c	K	M	ρ	P0	WIPq	CTq	X
1	3	12	2	10	10	0.852	0.064	1.45	0.071	20.50
2	3	12	3	10	10	0.645	0.098	0.32	0.013	23.22
3	3	12	4	10	10	0.496	0.105	0.062	0.0026	23.84

Table 10.

Model	c	WIP	CT	CTqq	WIPqq
1	2	3.16	0.15	0.102	2.46
2	3	2.25	0.09	0.046	1.67
3	4	2.05	0.085	0.028	1.36

Lambda effective (λ_e) for model is 1: 3.165/0.154 = 20.50; now enter the following:

```
names(y4)
[1] "Inputs"        "RO"          "Lq"          "VNq"          "Wq"
[6] "Throughput"    "L"           "VN"           "W"           "Lqq"
[11] "Wqq"          "Pn"          "Qn"
```

To see what these represent, for example, one can use the *R$* command:

```
y4$Pn
[1] 1.059244e-01 2.648109e-01 2.979123e-01 1.986082e-01
8.689109e-02
[6] 3.258416e-02 1.018255e-02 2.545637e-03 4.773070e-04
5.966338e-05
[11] 3.728961e-06
```

Returns the mean number of customers in queue when there is queue in the M/M/c/K/m queueing model "Pn" returns the system probabilities of the queueing model; for example the probability that the system is empty is P0 =1.059244e-01.

Exercises

Let us consider an M/M/1 station where 3 types of pieces are processed. The type 1 ones have the highest priority, while the type 3 ones have the lowest priority. Let us assume that we have the operating data of Table 11.

Get the cycle time as well as the amount of work in process for each piece for the two types of prioritization: preemptive and non-preemptive. Let us assume that the customer interarrival time follows an exponential distribution.

1. For the above exercise, the waiting cost for the pieces is given in the Table 12.

Table 11. Cycle time vs. batch size

Class	Service Time	Variance	Demand (Pieces/min)
1	1 min.	1	0.75
2	1.5 min.	0.25	1
3	0.75 min.	0.15	0.5

Table 12. Waiting cost

Type	CWIP
1	3.5
2	2.5
3	0.75

The station's operating cost is $4.5. Estimate the total operating cost of the work station for every type of priority rule. Let us assume that the cost expression is

$$TotalCost = Wipc_{Wait} + ServersC_{Server}$$

2. There are banks that regularly operate with a service approach in which the customers are classified into VIP and non-VIP. The operating data of one branch are as shown in Table 13.

There are 3 tellers in the branch. Estimate for each class of customer the queue cycle time and the number of customers waiting, assuming a non-preemptive prioritization, increase the value of x by 0.5 units (that represents the demand of the non-VIP customers), starting at 2.0 and ending up at 5 customers. Add servers when required.

Table 13. Service time, variability, and arrival rate

Type	Service Time	Variance	Arrival Rate (Client/hr)
VIP	10 min.	1	4
Non-VIP	10 min.	1	x

3. Let us consider an M/G/1 work station where two pieces are processed. The mean is 1.35 minutes for class 1 and 0.5 minutes for class 2. 1 piece arrives at the station every 3 minutes. The proportion of type 1 pieces is 10%. Estimate constant C of Kleinrock's conservation law.
4. Let us consider a work station where a process is done by batches. The average service time is 14.07 minutes and $C^2_s = 2$. The pieces are sent from another section of the plant at the rate of 1 every quarter of an hour. The dispatches are highly irregular as the estimated variability of the interarrival times is $C^2_a = 2$.
    a. Construct a spreadsheet that enables you to assess the cycle time for different values of β.
    b. Estimate the batch size that minimizes the cycle time.
    c. Assume that the variability of the interarrival time is reduced to $C^2_a = 1$. What is the new value for the cycle time?

REFERENCES

Adan, I., & Resing, J. (2001). *Queueing Theory*. Retrieved from http://www.win.tue.nl/~iadan/queueing.pdf

Atar, R., Giat, Ch., & Shimkin, N. (2010). The cμ/θ rule for many server queues with abandonment. *Operations Research*, *58*(5), 1427–1439. doi:10.1287/opre.1100.0826

Australasian College for Emergency Medicine. (2016). *Guidelines on the implementation of Australasian Triage Scale in Emergency Department*. Retrieved from https://acem.org.au/getattachment/4320524e-ad60-4e7c-a96d-bdf90cd7966c/G24-Implementation-of-the-Australasian-Triage-Scal.aspx

Bolch, G., Greiner, S., de Meer, H., & Trivedi, K. S. (2006). *Queueing networks and Markov chains*. Wiley. doi:10.1002/0471791571

Bondi, A. B., & Bunzen, J. P. (1981). The response times of priority classes under preemptive resume in M/G/m queues. *Operations Research*, *31*(3), 456–465.

Burke, P. J. (1975). Delays in single server queues with batch input. *Operations Research*, *23*(4), 830–833. doi:10.1287/opre.23.4.830

Chauhan, M. S., & Sharma, G. C. (1996). Profit analysis of M/M/r queueing model with balking and reneging. *MonteCarlo Methods and Aplications*, *2*(2), 139–144.

Cobham, A. (1954). Priority Assignments in Waiting Line Problems. *Operations Research*, *2*(1), 70–76.

Cooper, R. B. (1981). *Introduction to Queueing Theory* (2nd ed.). New York: Elsevier North Holland.

Curry, G. L., & Feldman, R. M. (2011). *Manufacturing systems modeling and analysis* (2nd ed.). Berlin: Springer. doi:10.1007/978-3-642-16618-1

Dai, J. G., & He, S. (2010). Customer abandonment in many servers queues. *Mathematics of Operations Research*, *35*(2), 347–362. doi:10.1287/moor.1100.0443

Hopp, W. J., & Spearman, M. L. (2008). *Factory physics*. Long Grove, IL: Waveland Press.

Isserson, K. V., & Moskop, J. C. (2007). Triage in Medicine, Part I: History and Types. *Annals of Emergency Medicine*, *49*(3), 275–287. doi:10.1016/j.annemergmed.2006.05.019 PMID:17141139

Kella, O., & Yechiali, U. (1985). Waiting Times in the Non-Preemptive Priority M/M/c Queue. *Communications in Statistics. Stochastic Models*, *1*(2), 257–262. doi:10.1080/15326348508807014

Kleinrock, L. (1965). A conservation law for a wide class of queueing disciplines. *Naval Research Logistics Quarterly*, *12*(2), 181–192. doi:10.1002/nav.3800120206

Lawrence, S. R., & Buss, A. H. (1995). Economic Analysis of Production Bottlenecks. *Mathematical Problems in Engineering*, *1*(4), 341–363. doi:10.1155/S1024123X95000202

Pilehvar, A. (2007). *Queieng networks approximations for mass dispensing and vaccination clinics* (Master Thesis). University of Maryland. Retrieved from http://drum.lib.umd.edu/bitstream/handle/1903/6927/umi-umd-4427.pdf?sequence=1

Stewart, W. J. (2009). *Probability, Markov Chains, Queues, and Simulation*. Princeton University Press.

Sztrick, J. (2012). Basic Queueing Theory. *Globe Editorial.* Retrieved from http://irh.inf.unideb.hu/~jsztrik/publications/books/GlobeEdit_Basic_Queueing_Theory_Sztrik_2016.pdf

Van der Heijden, M., van Harten, A., & Sleptchenko, A. (2004). Approximations for Markovian multi-class queues with preemptive priorities. *Operations Research Letters, 32*(3), 273–282. doi:10.1016/j.orl.2003.09.001

Wagner, H. (1969). *Principles of Operations Research.* Prentice Hall.

Ward, A. R., & Glynn, P. L. (2005). A difussion approximation for a GI/GI/1 queue with balking and reneging. *Queueing Systems, 50*(4), 371–400. doi:10.1007/s11134-005-3282-3

White, J. A., Schmidt, J. W., & Bennett, G. K. (1975). *Analysis of Queueing Systems.* London: Academic Press.

Whitt, W. (2006). Fluid models for multiserver queues with abandonment. *Operations Research, 54*(1), 37–54. doi:10.1287/opre.1050.0227

Williams, T. M. (1980). Nonpreemptive Multi-Server Priority Queues. *The Journal of the Operational Research Society, 31*(12), 1105–1007. doi:10.1057/jors.1980.206

Wu, K., Leon, F. M., & Zwart, B. (2011). Approximating the Performance of a Batch Service Queue Using the M/M[x]/1 model. *IEEE Transactions on Automation Science and Engineering, 8*(1), 95–102. doi:10.1109/TASE.2010.2041927

Chapter 8
Software for Queueing Analysis

ABSTRACT

Chapter 8 gives a brief discussion of computer simulation for discrete events. The chapter lists software programs in the technical literature that outline programs for the simulation of discrete events, both of commercial origin and free programs. In addition to the lists submitted, the authors present specialized packages for analysis and simulation of waiting lines in the R language. Statistical considerations are presented, which must be taken into account when obtaining data from simulations in situations of waiting lines. Chapter 8 presents three packages of the statistical program R: the "queueing" analysis package provides versatile tools for analysis of birth- and death-based Markovian queueing models and single and multiclass product-form queueing networks; "simmer" package is a process-oriented and trajectory-based discrete-event simulation (DES) package for R; and, the purpose of the "queuecomputer" package is to calculate, deterministically, the outputs of a queueing network, given the arrival and service times of all the customers. It also uses simulation for the implementation of a method for the calculation of queues with arbitrary arrival and service times. For each theme, the authors show the use of the packages in R.

COMPUTER SIMULATION

Computer simulation is the production of data about the behavior of a system or process, using an algorithm or model that reproduces the characteristics that are of interest. This algorithm or model is programmed in computer where the data of the behavior under study is also collected.

DOI: 10.4018/978-1-5225-5264-2.ch008

Computer simulation is used to model systems or processes with practical difficulties to be studied in a direct analytical way. To perform a computer simulation, a model or algorithm is programmed that reproduces the characteristics that are of interest in the study of the system or process. When executing the simulation program, we obtain data that allows us to understand the behavior that is reproduced in the system or process. The data obtained serve to carry out the necessary analyzes.

Computer simulation uses a programmed algorithm that represents the behavior of a system over time. The programmed algorithm uses a model with logical and mathematical relations of the elements that are identified as important to explain the observed behavior of the system under study.

The model or algorithm constructed to reproduce the behavior under study is validated. This implies, among other tests, that a previously observed behavior is reproduced with known conditions. The model must also allow us to obtain useful data for the analyzes that are planned to be carried out.

In queueing systems, the evolution of the system in time depends on the interactions of the timing of various discrete events. The state of such dynamic systems changes only at these discrete instants of time instead of continuously. We shall call such systems discrete event dynamic systems.

We can visualize such a discrete event dynamic system consisting of jobs and resources. Jobs travel from resource to resource demanding and competing for service. The dynamics of the system is determined by the interactions of the timings of various discrete events associated with the jobs and resources.

Discrete event simulation utilizes a mathematical and logical model of a system that portrays state changes at precise points in simulated time. Both the nature of the state changes and the time at which the changes occur mandate precise description.

Discrete event simulation languages provide the constructs of jobs, resources, timing of events. and logical tests while the coding in such languages produces the description of the specific systems.

These specialized programs allow the person who wants to perform the simulation of a waiting line system to be able to concentrate on the details that define the features that make the system under study of special interest.

Most of the efforts in simulation have to do with the following:

1. Good language design and friendly software to make simulation modeling easy;
2. Statistical analysis of outputs,

Software Lists

Every two years the magazine "OR/MS Today" has published reviews of software for discrete event simulation, (Swain, 2013) (Swain, 2015), and keeps a site on the network for consulting the most up-to-date results of the review. The document warns that the review should not be taken as complete but is representative of the available simulation packages. The original article presents 55 products from 31 vendors.

Using the results of Swain, Dias *et al.* as a basis, in 2016 they published an article that presents 19 commercial simulation programs based on popularity. This popularity is a measure made up by a combination of several measurements that include instances on the world-wide web and in scientific publications. The authors state that popularity does not ensure better quality or that any one program is a better simulation tool. However, there could be a positive correlation between popularity, quality and a better use of simulation. The authors say that the list was created using a subjective evaluation of a set of parameters. If, alternatively, we ere to use different parameters with different weightings we could get different results. Furthermore, the values of the parameters change over time and the evaluation needs to be updated to reflect these changes.

Furthermore, the University of Windsor professor, Myron Hlynka, maintains a list of queueing software, containing 73 products. (Hylinka, 2014).

A review of free discrete event simulation software is reported by Dagkakis and Heavey in 2016. However, the discussion of free software omits the discrete event simulation packages in the R statistical programming language: "simmer" and "queuecomputer".

Statistical Analysis of Simulation

Special consideration should be given to the statistical analysis of the results of a simulation of queues. The result of an experiment in queueing simulation is expected to be asymptotically valid. In other words, the measurements of interest to be obtained from the simulation make it possible to make valid inferences through large-scale simulation runs.

In queueing simulation, random arrival and service times are generated in accordance with the probability distributions that are believed best represent the behavior of the times; the interactions, in the form of arrival and service events, are calculated for random combinations of times, so the numerical results are different every time.

The expectation is that, in the long term after large numbers of these measurements have been observed and calculated, their distribution will tend to stabilize and there may be smaller variances in the mean measurements as observations increase.

For the variation in the means of the measurements to drop as the number of observations grows, certain conditions need to be fulfilled. One important condition is for the measurements observed to be independent of each other, in other words, for the value of the measurement observed in this moment not to be influenced by the values of previous measurements. In queues, we sometimes find that if we are measuring something such as a delay in a customer being served, this measurement can be strongly affected by a delay in the case of the previous customer. Thus, the measurement of the present customer is affected by the measurement of the previous customer, so it is not possible to ensure independence.

The measurements taken are also expected to maintain the same distribution. Some queueing systems have dynamic service rules. For example, in the case of blockage, when the capacity for "k" number of customers in the system is exceeded, no more arrivals can be permitted, and the system stays blocked until the number of customers in the system drops. This results in the distribution of arrival times being altered.

Law, in 2015, sums this up in three critical pitfalls, i.e., what could go wrong when analyzing the result of a simulation: 1. Calculating means and variances of the measurements under the assumption that they are independent of each other, when that is not the case. 2. When analyzing simulation experiments that do not have a set end point, not first letting the system run through its initial transitional stages and waiting for a run-time with stabilized behavior. 3. Failing to replicate the simulation experiment several times, thus making it impossible to get a sample of equally distributed independent measurements that the calculations can be based on with planned statistical precision.

Example 8.1: For example, if, in the simulation of a M/M/1 queueing system, we are interested in the mean time of customers in the system, assuming that the arrival speed is 2 customers an hour; the server's speed is 4 customers an hour and the arrival and service times are exponential, we will do 100 simulation replications resulting in:

```
0.4389063, 0.5226563, 0.4732985, 0.4779702, 0.5042095,
0.4513330, 0.5557048, 0.4579645, 0.4556518, 0.5007506,
0.4451871, 0.4440559, 0.4759613, 0.4579281, 0.4887823,
0.5306220, 0.4608935, 0.4905587, 0.4207742, 0.5350101,
0.4581040, 0.4783814, 0.4991897, 0.4556204, 0.4996850,
0.4409697, 0.5360867, 0.5283267, 0.5024605, 0.4923768,
0.4978618, 0.6024582, 0.5304764, 0.5256102, 0.5636591,
0.4331651, 0.4808827, 0.5677307, 0.5632884, 0.5069509,
0.4972176, 0.4617486, 0.5510866, 0.4885712, 0.5497090,
0.4989802, 0.4941889, 0.5016230, 0.6097744, 0.4849823,
0.5220226, 0.4730142, 0.4931017, 0.5467960, 0.4375041,
0.4341849, 0.4597312, 0.4662094, 0.4907573, 0.5097963,
0.5664588, 0.4837264, 0.5362580, 0.5564441, 0.5361276,
0.4085032, 0.4885115, 0.4712255, 0.5081124, 0.5390276,
0.5172462, 0.4493228, 0.4577551, 0.4611394, 0.5175619,
0.5152246, 0.4545553, 0.5584194, 0.4670837, 0.4599066,
0.4592233, 0.5475996, 0.5090994, 0.4722382, 0.5392428,
0.5200255, 0.4584708, 0.4719902, 0.5429865, 0.4745205,
0.5464634, 0.5767155, 0.4534279, 0.5192920, 0.5070653,
0.4975521, 0.5815280, 0.4537179, 0.4943039, 0.4886098
```

Being the result of random simulation experiments, the values are different. If the reported averages are based on just a few observations, the time in the system for each customer will not necessarily be able to be stabilized.

What we do get is that the replications are indeed independent of each other, as none of their values depend on the value of the others. As for their distribution: they have an mean of 0.4978919 and a standard deviation of 0.04187241. If an Anderson – Darling test for normality is applied we get a value P of 0.6106, so they can be considered to follow a normal distribution. A confidence interval 1-α with distribution t can be constructed for the mean time within the system using the formula:

$$\overline{x} \pm t_{n-1,1-\alpha/2} \frac{s}{\sqrt{n}} \quad (1)$$

In equation (1), the quantity *x* is added to and subtracted from the mean of the observations to get the upper and lower bounds for the confidence level, the value of the distribution *t* (quantile) for *n*-1 degrees of freedom; *n* is the number of observations, with a confidence level 1-$\alpha/2$, multiplied by the standard deviation of the data and divided by the square root of n.

If α is taken as being 0.05, then quantile t is 0.975 and 99 degrees of freedom corresponds to 1.984217.

The confidence bounds based on the 100 replications will be for a confidence level of 0.95 (α =0.05): lower bound =0.4895835 and upper bound= 0.5062002

In this way the replication of the experiments helps us to have independent estimations while the calculation of the confidence interval gives us a vision of the precision achieved by simulation.

R Language

R is a programming language for probability and statistical analysis that enables us to manipulate, calculate and plot graphs with data. For this purpose, it has developed a collection of built-in data analytical tools, using the programming language called "S" and making it possible to perform matrix and vector calculations. (R Core Team, 2016)

R is designed to be a quick way of developing new methods of interactive data analysis. It has thus grown by means of a large collection of "packages".

All the functions and datasets in R are saved in the packages. A package's contents become available when you download the package.

Standard Packages

There is a group of standard or base packages that form the start of the R code, containing the basic functions that make it possible for R to operate and the datasets as well as the graphic and basic statistical functions. They are available in the initial installation of R. The list of packages with which R is distributed are:

- **Base:** Base R functions.
- **Compiler:** R Bytecode compiler.
- **Datasets:** Base R data sets.
- **grDevices:** Graphic devices for base and grid graphics.
- **Graphics:** R functions for base graphics.
- **Grid:** A re-write of the graphic layout capabilities as well as some interaction support.
- **Methods:** Formally-defined methods and classes for R objects, plus other programming tools as described in the Green Book (Chambers, 1998).

- **Parallel:** Support for parallel computation, including by forking and by sockets, and random-number generation.
- **Splines:** Regression spline functions and classes
- **Stats:** R statistical functions.
- **Stats4:** Statistical functions using S4 classes.
- **Tcltk:** Interface and language links to Tcl / Tk GUI elements.
- **Tools:** Tools for package development and administration.
- **Utils:** R utility functions

Recommended Packages

Apart from the standard packages, the installation of R also brings another group of packages that are called "recommended". The list is as follows:

- **KernSmooth:** Functions for kernel smoothing (and density estimation) corresponding to the book "Kernel Smoothing" by M. P. Wand and M. C. Jones, 1995.
- **MASS:** Functions and datasets from the main package of (Venables &Ripley, 2013).
- **Matrix:** A Matrix package.
- **Boot:** Functions and datasets for bootstrapping from the book "Bootstrap Methods and Their Applications" (Davison & Hinkley, 1997).
- **Class:** Functions for classification (k-nearest neighbor and LVQ - *Learning Vector Quantization*).
- **Cluster:** Functions for cluster analysis.
- **Codetools:** Code analysis tools.
- **Foreign:** Functions for importing and exporting data stored by statistical software.
- **Lattice:** Lattice graphics, an implementation of Trellis Graphics functions.
- **Mgcv:** Routines for GAMs - Generalized Additive Models - and other generalized ridge regression problems with multiple smoothing parameter selection by GCV- Generalized Cross Validation- or UBRE - Un-Biased Risk Estimator.
- **Nlme:** Fit and compare Gaussian linear and nonlinear mixed-effects models.
- **Nnet:** Software for single hidden layer perceptrons ("feed-forward neural networks"), and for multinomial log-linear models.

- **Rpart:** Recursive partitioning and regression trees.
- **Spatial:** Functions for kriging and point pattern analysis (Venables & Ripley, 2013).
- **Survival:** Functions for survival analysis, including penalized likelihood.

Contributed Packages

As we have already mentioned, the standard and recommended packages include the R program in all the installations. But there also are thousands of additional contributed packages, written by many different authors. In general, the purposes of these packages is to implement specialist statistical methods, give access to data or hardware, and others are supplements for textbooks. The majority can be downloaded from "The Comprehensive R Archive Network" – CRAN – (https://CRAN.R-project.org/ and its mirrors) and other places such as Bioconductor (https://www.bioconductor.org/). and Omegahat (http://www.omegahat.net/). Many of these packages are classified in CRAN Task Views, making it possible to explore packages according to topic and providing tools for the automatic installation of all the packages for special areas of interest.

Package Management

In order to use an R package, the package must be installed and uploaded. A package needs to be installed before it can be loaded and its contents used. To see what packages are installed in its place, use the command without arguments:

```
> library()
```

If the user is connected to the Internet they can use the install.packages ("package name") command or the installation functions available in the Packages menu in the version of R for Windows. For example:

```
> install.packages("queueing")
```

If you want to load a package of interest, use the command library() but now with the argument equal to the requested package name, for example:

```
> library(queueing)
```

To see what packages are loaded, we can use the commands:

```
> search()
```

or

```
sessionInfo()
```

If you want to see the help topics in an installed package, you can use:

```
> help.start()
```

To start a help system session in HTML and then look through the list of help section packages.

As we have already mentioned, in 2016 Dagkakis and Heavey reviewed free software for Discrete Event Simulation. However, the discussion of free software omits the discrete event simulation packages in the R statistical programming language that is available in the CRAN package repositories: "simmer" and "queuecomputer".

There is also an R package called "queueing" that makes direct calculations by formula.

Table 1 presents some of the basic features of interest in this free software.

In this part we shall focus on the R packages available in the CRAN package repository that can be used for queueing analysis in the R statistical programming language.

Queueing Package

The queueing package (Canadilla, 2016) enables the analysis of Markov queueing models and queueing networks, handling the following models:

```
M/M/1, M/M/c, M/M/∞, M/M/1/K, M/M/c/K, M/M/c/c, M/M/1/K/K,
M/M/c/K/K, M/M/c/K/m, M/M/∞/K/K, Multiple Channel Open Jackson
Networks, Multiple Channel Closed Jackson Networks, Single
Channel Multiple Class Open Networks, Single Channel Multiple
Class Closed Networks and Single Channel Multiple Class Mixed
Networks. Also it provides a B-Erlang, C-Erlang and Engset
calculators.
```

Table 1. Characteristics of three packages for queueing theory

	queueing	simmer	queuecomputer
1 Language	R (>= 2.11.1)	core written in C++ R (>= 3.1.2)	R (>= 3.3.1)
2 OS Licence	GPL-2	MIT	GPL (>= 2)
3 Latest Stable release	0.2.10 /2017-05-29	3.6.2 / 2017-05-30	/ 2017-04-11
4 Latest Update	2017-05-29	2017-05-30	2017-04-11
5 Domain	analysis of birth and death based Markovian Queueing Models and Single and Multiclass Product-Form Queueing Networks *	A process-oriented and trajectory-based Discrete-Event Simulation (DES)	simulating queues with arbitrary arrival and service times, of general form G(t)/G(t)/K(t)/∞/n/FCFS
6 Simulation Technique	No	Discrete Event Simulation	Algorithm for queue departure computation (QDC)
7 Documentation	Reference manual	Reference manual, 7 vignettes	Reference manual, 2 vignettes
8 Communication	https://cran.r-project.org	http://r-simmer.org; https://cran.r-project.org	https://github.com/AnthonyEbert/queuecomputer; https://github.com/AnthonyEbert/queuecomputer
9 Version Control	yes	yes	yes

(*) queueing does not make networks with M/M/c/k nodes

For its installation, if you have an Internet connection, use the command:

```
> install.packages("queueing")
```

The program asks you to indicate the site where you can download the package, indicates the one it considers is better, then after the package is installed, it is loaded with the command:

```
> library(queueing)
```

The Use of Models in Queueing Package

All models are used in the same way:

- Create inputs calling the appropiate NewInput.model. For example:

```
x <- NewInput.MM1(lambda=6, mu=12, n=0)
```

for a M/M/1 model. To know the exact acronym model to use for NewInput function, you can search the html help or write help.search("NewInput") at the command line.

- Solve the model calling y <- QueueingModel(x). In this step, the CheckInput(x) will be called, to check the parameters of the queueing model.
- Finally, you can get a performance value as W(y), the mean time spend in a queueing system, or a report of the principals performance values calling summary(y)

Example 8.2: For a M/M/1 System, with an arrival rate of 6 and service rate of 12, a single server plus infinite capacity, we get:

First the input object, i_mm1, is created, allocating the command

```
NewInput.MM1(lambda=6, mu=12, n=0),
```

with the input values; n=0. This is so that only the probability of there being zero units in the system is shown. If we want to see more probability values, the value of n is adjusted as required.

```
> i_mm1 <- NewInput.MM1(lambda=6, mu=12, n=0) #input
```

An output object, o_mm1 is created, with the M/M/1 model, using the command QueueingModel(i_mm1):

```
> o_mm1 <- QueueingModel(i_mm1)
```

To see the results, use the command summary(o_mm1):

```
> summary(o_mm1)
The inputs of the M/M/1 model are: lambda: 6, mu: 12, n: 0
The outputs of the M/M/1 model are:
The probability (p0, p1,..., pn) of the n = 0 clients in the
system are: 0.5
The traffic intensity is: 0.5
The server use is: 0.5
The mean number of clients in the system is: 1
The mean number of clients in the queue is: 0.5
The mean number of clients in the server is: 0.5
The mean time spend in the system is: 0.166666666666667
```

```
The mean time spend in the queue is: 0.0833333333333333
The mean time spend in the server is: 0.0833333333333333
The mean time spend in the queue when there is queue is:
0.166666666666667
The throughput is: 6
The calculation of specific performance measurements can also
be requested:
> Lqq(o_mm1) # the mean number of customers in queue when there
is queue in a queueing model
[1] 2
> Wqq(o_mm1) # the mean time spend in queue when there is queue
in a queueing model
[1] 0.1666667
> L(o_mm1) # the mean number of customers in a queueing model
[1] 1
> W(o_mm1) # the mean time spend in a queueing system
[1] 0.1666667
> Lq(o_mm1) # the mean number of customers in the queue in a
queueing model
[1] 0.5
> Wq(o_mm1)# the mean time spend in queue in a queueing model
[1] 0.08333333
> VN(o_mm1)# the variance of the number of customers in a
queueing model
[1] 2
> VT(o_mm1)# Returns the variance of the time spend in a
queueing model
[1] 0.02777778
> VNq(o_mm1)# Returns the variance of the number of customers
in the queue in a queueing model
[1] 1.25
> VTq(o_mm1)# Returns the variance of the time spend in queue
in a queueing model
[1] 0.02083333
```

Example 8.3: An analysis is being made of a queuing system with several servers. Start with an arrival rate (λ) of 20 units an hour and c= 3 servers, each with a service rate (μ) equal to 8 units an hour. Performance measurements are required.

First, enter the model, with the given parameters, method = 0 means that you want the exact calculation, n=10 means that you want to calculate the probabilities for 0, 1, 2, …, 10 customers:

```
> i_mmc <- NewInput.MMC(lambda=20, mu=8, c=3, n=10, method=0)
```

The model is then checked:

```
> CheckInput(i_mmc)
```

There is no warning about the model. It is considered to be correct and an object is now obtained with the performance measurements:

```
> o_mmc<-QueueingModel(i_mmc)
```

Get the summary of the performance measurements for the model:

```
> summary(o_mmc)
The inputs of the model M/M/c are: lambda: 20, mu: 8, c: 3, n:
10, method: Exact
The outputs of the model M/M/c are:
The probability (p0, p1,..., pn) of the n = 10 clients in the
system are:
0.04494382 0.1123596 0.1404494 0.1170412 0.09753433 0.08127861
0.06773218 0.05644348 0.04703623 0.03919686 0.03266405
The traffic intensity is: 2.5
The server use is: 0.833333333333333
The mean number of clients in the system is: 6.01123595505618
The mean number of clients in the queue is: 3.51123595505618
The mean number of clients in the server is: 2.5
The mean time spend in the system is: 0.300561797752809
The mean time spend in the queue is: 0.175561797752809
The mean time spend in the server is: 0.125
The mean time spend in the queue when there is queue is: 0.25
The throughput is: 20
```

Comparing Models

You might want to compare other queueing system configurations, for example, you can compare alternatives by increasing the number of servers or decreasing service time. The objects for the proposed models are generated by using the appropriate commands:

```
> y1 <- QueueingModel(NewInput.MM1(lambda=20, mu=25))
> y2 <- QueueingModel(NewInput.MMC(lambda=20, mu=8, c=3))
> y3 <- QueueingModel(NewInput.MMInf(lambda=20, mu=4))
> y4 <- QueueingModel(NewInput.MMCC(lambda=20, mu=8, c=4))
```

The performance measurements are put in tabular form for the four models This allows us to compare them:

```
> CompareQueueingModels(y1, y2, y3, y4)
```

Obtaining:

```
   lambda  mu  c  k  m  RO   PO     Lq  Wq   Throughput  L    W
Wqq  Lqq
1  20      25  1  NA NA 0.8  0.2    3.2 0.16 20.00           4.00
0.20 0.20 5
2  20      8   3  NA NA 0.83 0.0449 3.5 0.17 20.00           6.01
0.30 0.25 6
3  20      4   NA NA NA 5    0.0067 0.0 0.0  20.00           5.00
0.25 NA   NA
4  20      8   4  4  NA 0.53 0.0921 0.0 0.0  17.00           2.12
0.12 NA   NA
```

Model 2 is the original condition of the system and “NA” means that there is available datum. Model 1 implies a single server with a speed of 25 units an hour, whereas model 3 is an alternative self-service server, assuming a service speed of 4 units an hour for every self-service unit. Model 4 implies a service limit with a maximum of 4 customers being served.

Example 8.4: Jackson Network - For example you have a network with the following arrival, service and configuration values: 35 transactions an hour arrive at node 1 which are attended with FIFO discipline by a single server at a rate of 120 transactions an hour; while 55% of the transactions are channeled to node 2, which has three servers that serve at a rate of 10 units an hour each, 2% of the transactions go to node 3, 45% of the transactions are received from node 1 in node 3 which has 7 servers that operate at a rate of 3 units an hour and 1% of the node 3’s transactions go on to node 2.

The matrix that represents the connectivity between the three nodes, ends up as:

```
      [,1]    [,2]    [,3]
[1,]  0       0.55    0.45
[2,]  0       0       0.02
[3,]  0       0.01    0
```

where the connectivities from i to node j, (i,j), are:

(1,2)= 0.55, (1,3)= .45, (2,3)= 0.02, (3,2)= 0.01.

The three nodes are defined:

```
> node1<-NewInput.MM1(lambda=35, mu=120, n=0)
> node2<-NewInput.MMC(lambda=0, mu=10, c=3, n=0, method=0)
> node3<-NewInput.MMC(lambda=0, mu=3, c=7, n=0, method=0)
```

The matrix is entered:

```
> m <- c(0, 0.55, 0.45, 0, 0, 0.02, 0, 0.01, 0)
> prob <- matrix(data=m, nrow=3, ncol=3, byrow=TRUE)
```

The object of a Jackson open network is formed:

```
> ojn1 <- NewInput.OJN(prob, node1, node2, node3)
```

The results object is obtained for the network:

```
> m_ojn1 <- QueueingModel(ojn1)
> summary(m_ojn1)
```

The inputs of the open Jackson network are:

```
$prob
 [,1] [,2] [,3]
[1,] 0 0.55 0.45
[2,] 0 0.00 0.02
[3,] 0 0.01 0.00
$nodes
$nodes[[1]]
$lambda
[1] 35
$mu
[1] 120
$n
[1] 0
attr(,"class")
[1] "i_MM1"
$nodes[[2]]
$lambda
[1] 0
$mu
[1] 10
$c
[1] 3
$n
```

```
[1] 0
$method
[1] 0
attr(,"class")
[1] "i_MMC"
$nodes[[3]]
$lambda
[1] 0
$mu
[1] 3
$c
[1] 7
$n
[1] 0
$method
[1] 0
attr(,"class")
[1] "i_MMC"
attr(,"class")
[1] "i_OJN"
The outputs of the open Jackson network are:
---------- Complete network -------------------------
The mean number of clients in the network is: 9.89862790169203
The mean time spend in the network is: 0.282817940048344
The throughput of the network is: 35
--------- Per node ---------------------------------
The use of node 1 is: 0.291666666666667
The throughput of node 1 is: 35
The mean number of clients in node 1 is: 0.411764705882353
The mean time spend in node 1 is: 0.0117647058823529
The probability (p0, p1,..., pn) or visit ratio of node 1 is:
[1] 0.7083333
The use of node 2 is: 0.647046075881843
The throughput of node 2 is: 19.4113822764553
The mean number of clients in node 2 is: 2.70592608787711
The mean time spend in node 2 is: 0.0773121739393459
The probability (p0, p1,..., pn) or visit ratio of node 2 is:
[1] 0.120788
The use of node 3 is: 0.768487030739481
The throughput of node 3 is: 16.1382276455291
The mean number of clients in node 3 is: 6.78093710793257
The mean time spend in node 3 is: 0.193741060226645
The probability (p0, p1,..., pn) or visit ratio of node 3 is:
[1] 0.003779239
```

Some commands in the queueing package are shown in Table 2.

Table 2. Commands of queueing package

Function	Argument
CheckInput(x,...) to check the parameters of a queueing model (or network)	x: a object of class i_MM1, i_MMC, i_MM1K, i_MMCK, i_MM1KK, i_MMCKK, i_MMCC, i_MMCKM, i_MMInfKK, i_MMInf, i_OJN
CompareQueueingModels(model,...) CompareQueueingModels2(models) Compare several queueing models in a tabulated format model	model, ...: a separated by comma list of queueing models obtained calling QueueingModel models: A list of queueing models obtained calling QueueingModel o_MM1, o_MMC, o_MMInf, o_MM1K, o_MMCK, o_MMCC, o_MM1KK, o_MMCKK, o_MMCKM, o_MMInfKK classes can be compared
B_erlang(c=1, u=0) Returns the probability that all servers are busy	c: numbers of servers u: rate of arrivals / rate of service
C_erlang(c=1, r=0) Returns the probability to wait in queue because all servers are busy	c: numbers of servers r: rate of arrivals / rate of service
Engset(k=1, c=0, r=0) Returns the probability of blocking in a finite source model	k: numbers of users c: numbers of servers r: rate of arrivals / rate of service
Inputs(x,...) Returns the inputs parameters of a already built queueing model (or network)	x: a object of class o_MM1, o_MMC, o_MM1K, o_MMCK, o_MM1KK, o_MMCKK, o_MMCC, o_MMCKM, o_MMInfKK, o_MMInf, o_OJN, o_MCON, o_MCCN, o_MCMN
L(x,...) Returns the mean number of customers in a queueing model (or network)	x: a object of class o_MM1, o_MMC, o_MM1K, o_MMCK, o_MM1KK, o_MMCKK, o_MMCC, o_MMCKM, o_MMInfKK, o_MMInf, o_OJN, o_MCON, o_MCCN, o_MCMN
Lc(x,...) Returns the vector with the mean number of customers of each class in a multiclass queueing network	x: a object of class o_MCON, o_MCCN, o_MCMN
Lck(x,...) Reports a matrix with the mean number of customers of class i in each node (server) j in a MultiClass Network	x: a object of class o_MCON, o_MCCN, o_MCMN
Lk(x,...) Returns the vector with the mean number of customers in each node (server) of a queueing network	x: a object of class o_OJN, o_CJN, o_MCON, o_MCCN, o_MCMN
Lq(x,...) Returns the mean number of customers in the queue in a queueing model	x: a object of class o_MM1, o_MMC, o_MM1K, o_MMCK, o_MM1KK, o_MMCKK, o_MMCC, o_MMCKM, o_MMInfKK, o_MMInf
Lqq(x,...) Returns the mean number of customers in queue when there is queue in a queueing model	x: a object of class o_MM1, o_MMC, o_MM1K, o_MMCK, o_MM1KK, o_MMCKK, o_MMCC, o_MMCKM, o_MMInfKK, o_MMInf
Pn(x,...) Pn returns the probabilities that a queueing model (or network) has n customers.	x: an object of class o_MM1, o_MMC, o_MM1K, o_MMCK, o_MM1KK, o_MMCKK, o_MMCC, o_MMCKM, o_MMInfKK, o_MMInf, o_OJN.
Qn(x,...) Qn returns the probabilities that an arrival that enters the system see n customers in it.	x: an object of class o_MM1, o_MMC, o_MM1K, o_MMCK, o_MM1KK, o_MMCKK, o_MMCC, o_MMCKM, o_MMInfKK, o_MMInf

continued on following page

Table 2. Continued

Function	Argument
QueueingModel(x,...) Generic S3 method to build a queueing model (or network)	x: an object of class i_MM1, i_MMC, i_MM1K, i_MMCK, i_MM1KK, i_MMCKK, i_MMCC, i_MMCKM, i_MMInfKK, i_MMInf, o_OJN, o_MCON
RO(x,...) Reports the server use of a queueing model	x: an object of class o_MM1, o_MMC, o_MM1K, o_MMCK, o_MM1KK, o_MMCKK, o_MMCC, o_MMCKM, o_MMInfKK, o_MMInf
ROck(x,...) Reports a matrix with the use of class i in each node (server) j in a MultiClass Queueing Network	x: an object of class o_MCON, o_MCCN, o_MCMN
ROk(x,...) Reports a vector with each node (server) use of a queueing network	x: an object of class o_OJN, o_CJN, o_MCON, o_MCCN, o_MCMN
SP(x,...) Returns the saturation point of a queueing model	x: an object of class o_MM1KK
summary(x,...) generates a report of the principals performance values of the queueing object received as parameter	x: an object of class o_MM1, o_MM1K, o_MM1KK, o_MMC, o_MMCC, o_MMCK, o_MMCKK, o_MMCKM, o_MMInf, o_ MMInfKK, o_OJN, o_CJN, o_MCCN, o_MCMN, o_MCON
Throughput(x,...) Returns the throughput of a queueing model (or network)	x: an object of class o_MM1, o_MMC, o_MM1K, o_MMCK, o_MM1KK, o_MMCKK, o_MMCC, o_MMCKM, o_MMInfKK, o_MMInf, o_OJN, o_CJN, o_MCON, o_MCCN, o_MCMN
Throughputc(x,...) Reports a vector with each class throughput in a multiclass queueing network	x: an object of class o_MCON, o_MCCN, o_MCMN
Throughputck(x,...) Reports a matrix with the throughput of class i in each node (server) j in a MultiClass Network	x: an object of class o_MCON, o_MCCN
Throughputcn(x,...) Returns a matrix with the Throughput from each class and every population of a Multi Class Closed Network	x: an object of class o_MCCN
Throughputk(x,...) Reports a vector with each node (server) throughput of a queueing network	x: a object of class o_OJN, o_CJN, o_MCON, o_MCCN, o_MCMN
Throughputn(x,...) Returns a vector with the each Throughput from 1 to the parameter n (population passed as input) of a Closed Network	x: an object of class o_CJN
VN(x,...) Returns the variance of the number of customers in a queueing model	x: an object of class o_MM1, o_MMC, o_MM1K, o_MMCK, o_MM1KK, o_MMCKK, o_MMCC, o_MMCKM, o_MMInfKK, o_MMInf
VNq(x,...) Returns the variance of the number of customers in the queue in a queueing model	x: an object of class o_MM1, o_MMC, o_MM1K, o_MMCK, o_MM1KK, o_MMCKK, o_MMCC, o_MMCKM, o_MMInfKK, o_MMInf
VT(x,...) Returns the variance of the time spend in a queueing model	x: an object of class o_MM1, o_MMC, o_MM1K, o_MMCK, o_MM1KK, o_MMCKK, o_MMCC, o_MMCKM, o_MMInfKK, o_MMInf

continued on following page

Table 2. Continued

Function	Argument
VTq(x,...) Returns the variance of the time spend in queue in a queueing model	x: an object of class o_MM1, o_MMC, o_MM1K, o_MMCK, o_MM1KK, o_MMCKK, o_MMCC, o_MMCKM, o_MMInfKK, o_MMInf
W(x,...) Returns the mean time spend in a queueing model (or network)	x: an object of class o_MM1, o_MMC, o_MM1K, o_MMCK, o_MM1KK, o_MMCKK, o_MMCC, o_MMCKM, o_MMInfKK, o_MMInf, o_OJN, o_MCON, o_MCCN, o_MCMN
Wc(x,...) Returns the vector with each class mean time spend on a multiclass queueing network	x: an object of class o_MCON, o_MCCN, o_MCMN
Wk(x,...) Returns the mean time spend in each node (or server) of a network	x: an object of class o_OJN, o_CJN, o_MCON, o_MCCN, o_MCMN
Wq(x,...) Returns the mean time spend in queue in a queueing model	x: an object of class o_MM1, o_MMC, o_MM1K, o_MMCK, o_MM1KK, o_MMCKK, o_MMCC, o_MMCKM, o_MMInfKK, o_MMInf
Wqq(x,...) Returns the mean time spend in queue when there is queue in a queueing model	x: an object of class o_MM1, o_MMC, o_MM1K, o_MMCK, o_MM1KK, o_MMCKK, o_MMCC, o_MMCKM, o_MMInfKK, o_MMInf

Simmer Package

The simmer package for R language is a process-oriented and trajectory-based discrete event simulation - (DES) – package; the trajectory is the path that entities of the same type take in the simulation model. (Ucar & Smeets, 2017)

The SimPy and SimJulia.jl packages were useful examples for this package. The simulation core is written in C++ with automatic monitoring capacities. You have an API in R that revolves around the concept of trajectory. (Ucar, Smeets, & Azcorra, 2017)

The syntax for declaring the trajectories depends on the R package called "magrittr". The "magrittr" package gives an operator to the syntax for the trajectories. The operator appears as: %>%, with which you can pipe a value forward into an expression or function call. Resulting in it being written as: *x %>%* f, instead of $f\,(x)$.

SIMMER is a package for discrete event simulation. SIMMER runs in the R programming language environment. There is another package called simmer.plot, associated with the simmer package, that provides plotting methods for the results obtained in simmer.

The packages are first downloaded from repositories. When the package is downloaded, the R program recognizes that the other packages are needed for the installation of the package and also downloads them. After the package and its complementary packages has been downloaded, it is installed. The R statistical programming language also automatically installs the necessary packages.

Example 8.5: An example is given below of how to use simmer in the simulation of a M/M/1 system, assuming an arrival speed of: lambda=3 and a service speed: mu=6. First of all, the two packages, simmer and simmer.plot, are loaded.

```
>library(simmer)
>library(simmer.plot)
```

The values of the arrival speed (lambda) and service speed (mu) are established.

```
> set.seed(1234)
> lambda=3
> mu=6
```

Now the trajectory of the arrivals within the system is constructed

```
> mm1.trajectory <- trajectory() %>%
+ seize("resource", amount=1) %>%
+ timeout(function() rexp(1, mu)) %>%
+ release("resource", amount=1)
```

A trajectory is defined as a coupling of activities that altogether make up the life of the arrivals in the system. As soon as the arrival is coupled to the trajectory, in general, the activities in the trajectory begin to be processed in the specified order and, eventually, it will leave the system.

The "timeout" function produces a random waiting time. The function can interact with other entities in the simulation, make decisions or even simply take a constant value.

Once the trajectory is defined, an environment where the resources and their order are declared is created as well as the (arrivals) generator with its random function for generating times in accordance with a probability distribution. The created trajectory stays within the defined environment

```
> mm1.env <- simmer() %>%
+ add_resource("resource", capacity=1, queue_size=Inf) %>%
+ add_generator("arrival", mm1.trajectory, function() rexp(1,
```

```
lambda)) %>%
+ run(until=2000)
```

The number of customers in the System obtained through the simulation of 2,000 customers are graphically compared with the theoretical value L (number of units in the system)

```
> rho=lambda/mu
> L<-rho/(1-rho)
> plot(mm1.env, "resources", "usage", "resource",
items="system")
+ geom_hline(yintercept=L)M/M/1)
```

Figure 1. Client number vs. time. The expected number in the system is 1.0.

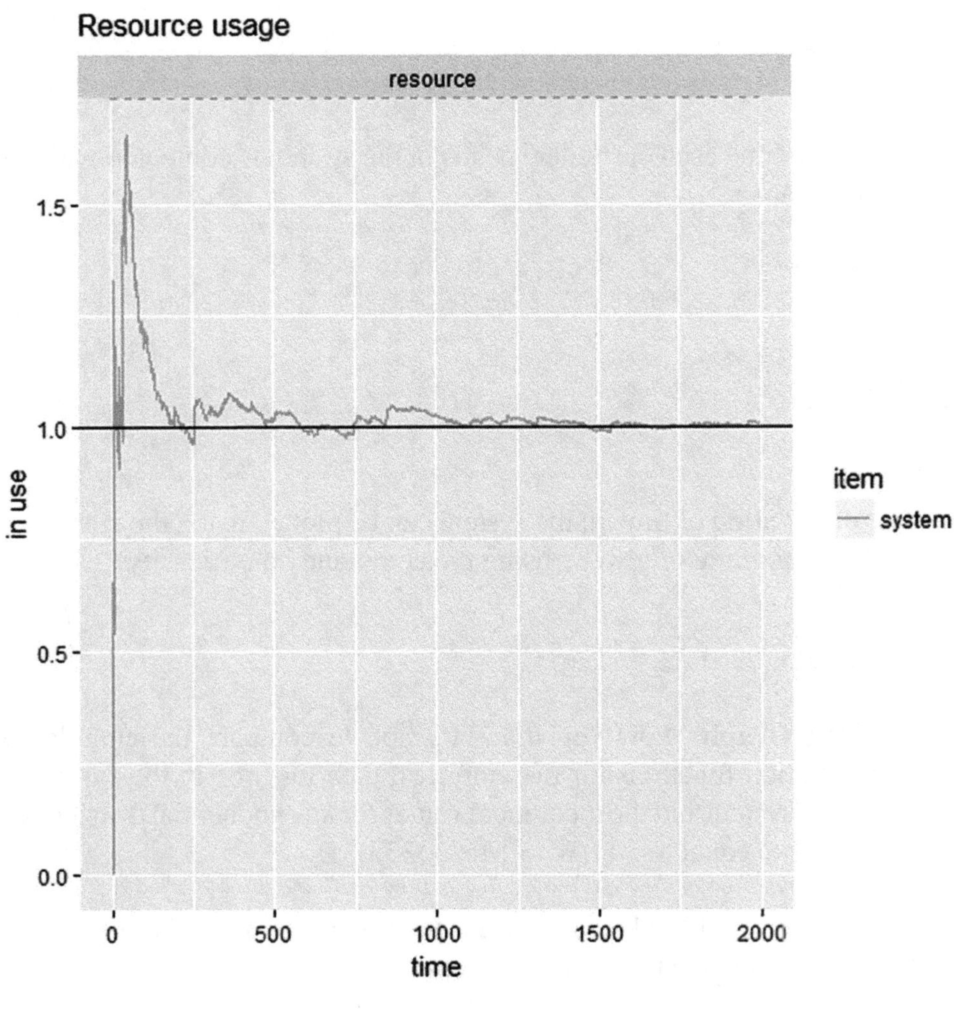

The inverse of the mean interarrival times is the rate observed in simulation which is compared with the lambda used for generating observations.

```
> lambda
[1] 3
> 1/mean(diff(subset(mm1.arrivals,finished==TRUE)$start_time))
[1] 3.043389
```

Resource usage can be plotted:

```
plot(mm1.env, "resources", "usage", "resource",
items=c("queue", "server"), steps=TRUE) +
 xlim(0, 20) + ylim(0, 4)
```

Warning messages:

```
1: Removed 24082 rows containing missing values (geom_path).
2: Removed 24082 rows containing missing values (geom_path).
```

The mean time each customer spends in the system is compared with the theoretical value:

```
> mm1.arrivals <- get_mon_arrivals(mm1.env)
> mm1.t_system <- mm1.arrivals$end_time - mm1.arrivals$start_
time
> W <- L / lambda
> W; mean(mm1.t_system)
[1] 0.3333333
[1] 0.3317824
```

The distribution of time in the system can be plotted using the command "hist" for histogram, or also by using the command "plot(density)":

```
> hist(mm1.t_system)
>plot(density(mm1.t_system))
```

- **ECDF Graph:** Now you draw a graph to compare the cumulative distribution functions for the empirical data for time in the system in mm1.t_system and the theoretical values for an exponential distribution with a rate equal to 3 (1/W = 1/0.333333)

t<-seq(0,4*0.3317824,0.1) # a sequence is generated of times between 0 and four times the mean of mm1.t_system.

Figure 2. Resource usage: Queue and server

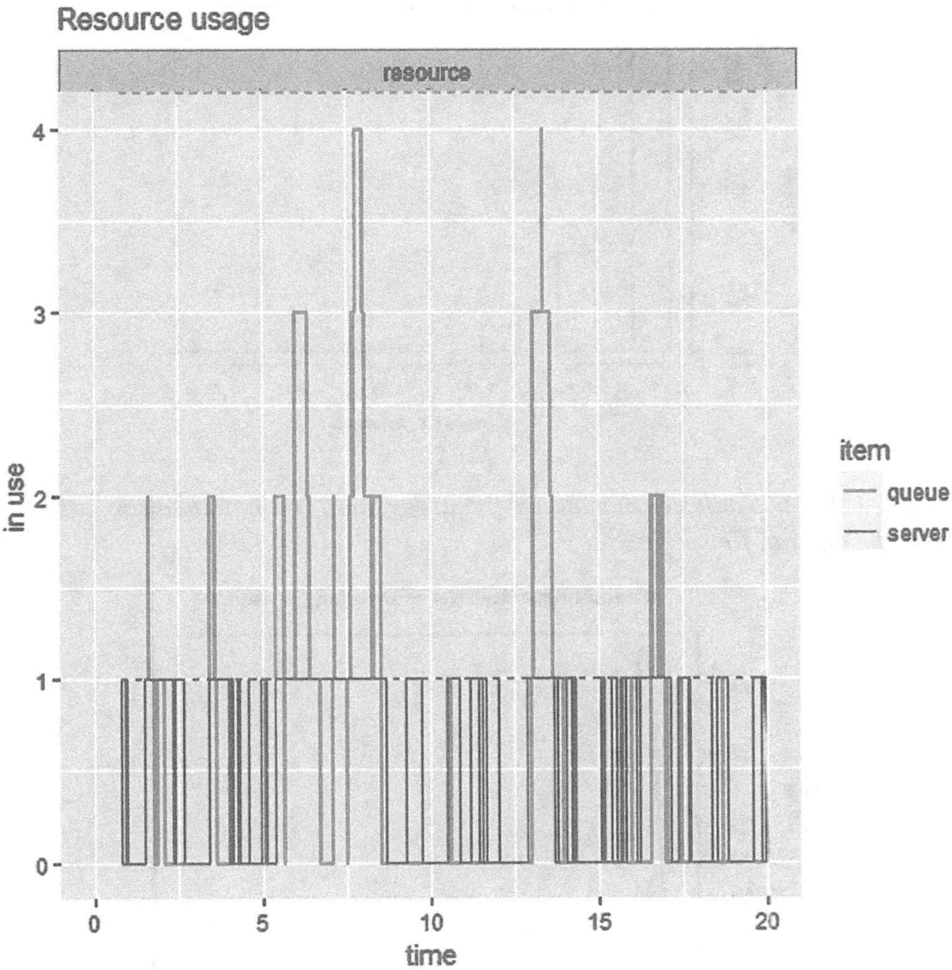


```
> plot(t,pexp(t,rate=3),type="l", main="ECDF and Exponential
CDF")
> plot(ecdf(mm1.t_system),add=TRUE)
```

- **Goodness of Fit Test:** A Kolmogorov-Smirnov test is run to compare the distribution of the times in system on vector mm1.t_system, with an exponential distribution with parameter 3:

```
> ks.test(mm1.t_system, "pexp", 3)
One-sample Kolmogorov-Smirnov test
data: mm1.t_system
```

Figure 3. Histogram (left) and density plot (right) of time in system

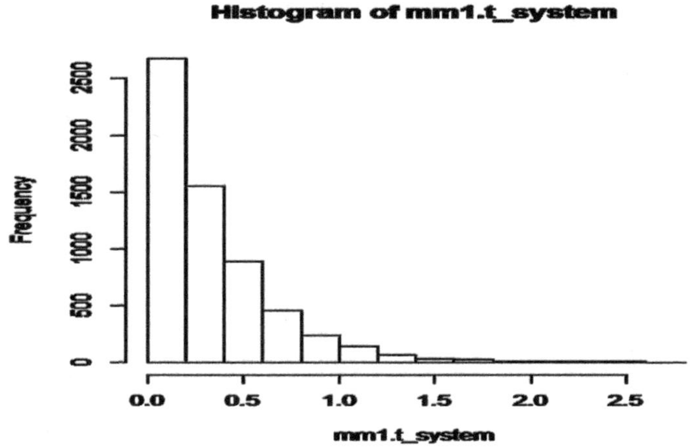

Figure 4. The two adjusted cumulative frequency lines. It is hard to distinguish one from the other at first sight.

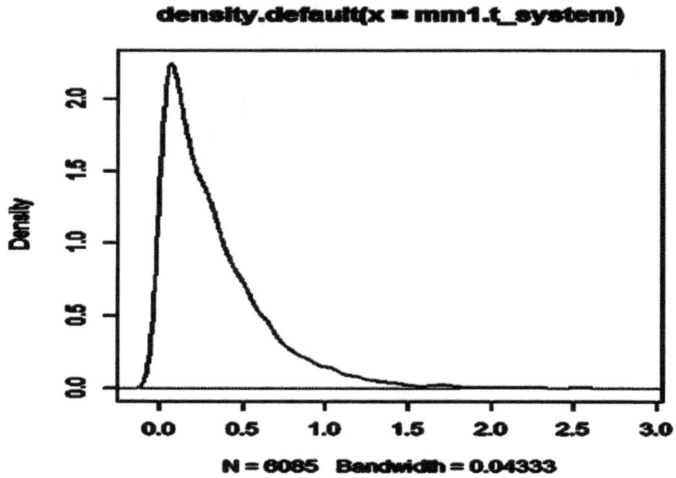


```
D = 0.016041, p-value = 0.08731
alternative hypothesis: two-sided
```

The value P is higher than 0.05 and the equality-of-distributions hypothesis holds.

Another way of testing probability distribution is by the Anderson-Darling test. In order to perform the Anderson-Darling test in the R statistical programming language, an additional package needs to be installed. Use the goftest package (Faraway, Marsaglia, Marsaglia & Baddeley, 2015). The instructions for its installation, once you have access to the web, are to give the command:

```
> install.packages("goftest")
```

A request will appear asking us to specify the site you want to download the package from. Once you have indicated the site and the command has been run, give the following command:

```
> library(goftest)
```

The syntax of the function for running the Anderson-Darling test has a similar structure to that of the Kolmogorov-Smirnov test: ad.test(x, "pdist", parameters), where x is the data vector you want to test, with the null hypothesis that it follows a distribution declared between inverted commas with the name of the cumulative probability distribution in R: "pdist", then with the necessary distribution parameters separated by commas.

```
>ad.test(mm1.t_system, "pexp",3)
 Anderson-Darling test of goodness-of-fit
 Null hypothesis: exponential distribution
data: mm1.t_system
An = 1.3976, p-value = 0.2029
```

The value P is higher than the significant values of 0.10, 0.05 and 0.01 that are commonly used, so the equality-of-distribution hypothesis for the empirical data with exponential distribution and parameter 3 holds.

- **Simulation Replications:** The simulation replications can be done by applying the command mclapply from the "parallel" package:

```
>library(parallel)
>envs <- mclapply(1:100, function(i) {
 simmer() %>%
 add_resource("resource", capacity=1, queue_size=Inf) %>%
 add_generator("arrival", mm1.trajectory, function() rexp(100,
lambda)) %>%
 run(1000/lambda) %>%
 wrap()
}, mc.set.seed=FALSE)
```

The command "wrap" is used in order not to lose the calculated values within the function. To get the mean times in the system, you have:

```
t_system <- get_mon_arrivals(envs) %>%
 dplyr::mutate(t_system = end_time - start_time) %>%
 dplyr::group_by(replication) %>%
 dplyr::summarise(mean = mean(t_system))
```

Resulting in 100 replications of the mean time in the System that are independent of each other. In other words, none of their values depend on those of the rest. Now, we must see if this sampling generated a dataset with identical distribution. We would expect a normal distribution of mean values. First, the mean and standard deviation of the 100 replications are calculated:

```
> mean(t_system$mean)
[1] 0.3335967
> sd(t_system$mean)
[1] 0.02855804
```

We can see the histogram with the following command

```
hist(t_system$mean)
```

The normality test is performed with the Anderson-Darling test, using the command ad.test() from the "goftest" package:

```
>library(goftest)
> ad.test(t_system$mean,"pnorm",0.3335967,0.02855804)
 Anderson-Darling test of goodness-of-fit
 Null hypothesis: Normal distribution
data: t_system$mean
An = 0.47578, p-value = 0.7712
```

The P value of 0.7712 indicates that the hypothesis that the data follow a normal distribution with mean = 1/6 and standard deviation =1/6 cannot be rejected.

Command t.test() is used to construct the confidence interval with 0.95 of the 100 generated replications:

```
>t.test(t_system$mean)
 One Sample t-test
data: t_system$mean
t = 116.81, df = 99, p-value < 2.2e-16
```

Figure 5. Histogram showing the observed distribution of the mean values of time in system for the 100 simulation replications.

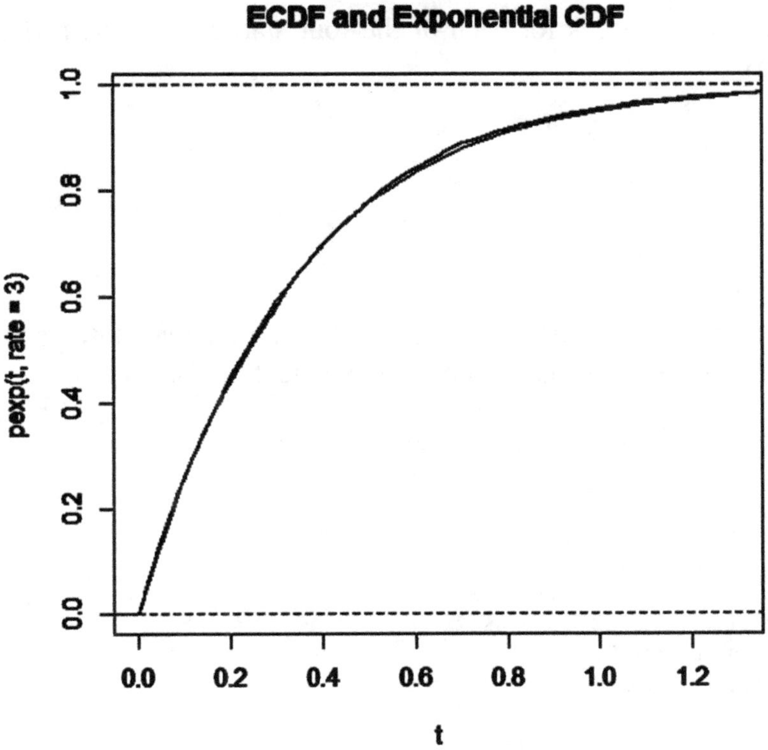


```
alternative hypothesis: true mean is not equal to 0
95 percent confidence interval:
 0.3279301 0.3392632
sample estimates:
mean of x
0.3335967
```

The confidence interval for the mean time in System is between a lower bound f 0.3279301 and an upper bound of 0.3392632 with a confidence level of 0.95 (alpha equal to 0.05. The replications make it possible to construct an interval that indicates the precision we can achieve in estimating a measurement of the system under the observed study.

Example 8.6: A Series System - The example is presented of four stations in series that process exponential arrivals in (also exponential) time. Each station has a limited queueing capacity k=2. In Kendall's notation the 4 stations are M/M/1/3. First load the simmer and simmer.plot packages:

```
library(simmer)
library(simmer.plot)
```

Now set the speeds for each of the four stations and the arrival speed (lambda).

```
mu1<-6
mu2<-12
mu3<-12
mu4<-6
lambda<-3
```

The trajectory that follows a product through the four stations is defined below, declaring a response time with exponential distribution:

```
product <- trajectory("products' path") %>%
seize("station1", 1) %>%
timeout(function() rexp(1, mu1)) %>%
release("station1", 1) %>%
seize("station2", 1) %>%
timeout(function() rexp(1, mu2)) %>%
release("station2", 1) %>%
seize("station3", 1) %>%
timeout(function() rexp(1, mu3)) %>%
release("station3", 1) %>%
seize("station4", 1) %>%
timeout(function() rexp(1, mu4)) %>%
release("station4", 1)
```

The environment is declared with the four resources with a M/M/1/3 model, the arrivals generator with exponential distribution is added and the request is made to run 1,000 events.

```
network.env <- simmer() %>%
add_resource("station1", capacity=1, queue_size=2) %>%
add_resource("station2", capacity=1, queue_size=2) %>%
add_resource("station3", capacity=1, queue_size=2) %>%
add_resource("station4", capacity=1, queue_size=2) %>%
add_generator("arrival", product, function() rexp(1, lambda))
%>%
run(until=1000)
```

The resource usage can be plotted with the plot command of simmer.plot:

```
plot(network.env, "resources", "usage", c("station1", "station
2","station3","station4"))
```

Queuecomputer Package

The purpose of the queuecomputer package is to calculate, deterministically, the outputs of a queuing network, given the arrival and service times of all the customers. Although it also uses simulation for the implementation of a method for the calculation of queues with arbitrary arrival and service times. It relies on the functions called: queue_step, lag_step and wait_step. (Ebert A., 2017)

The package is modular and is integrated with the R package dplyr. Complex queueing networks with series, parallel and fork/join topologies can be constructed using both packages. (Ebert, A., Wu, P., Mengersen, K., & Ruggeri, F., 2017).

`queue_step Function`

Figure 6. Comparison of resource usage in the four stations. The highest line indicates system usage, the lowest is the queue and the one in the middle is server usage.

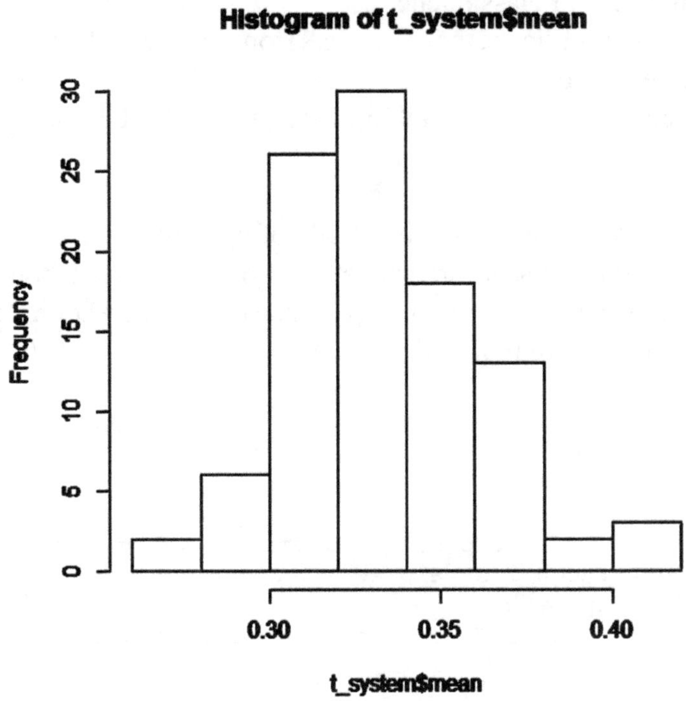

Based on the arrival and service times, the queue_step function calculates departure times and queue lengths for a queuing system.

```
queue_step(arrivals, service, servers= 1, labels= NULL, adjust=
1)
```

Where:

"arrivals" is a numeric vector of non-negative arrival times.

"service" is a numeric vector of service times with the same ordering as "arrivals".

"servers" is a a non-zero natural number that can also be an object of class "server.stepfun" or an object of class "server.list".

"labels" is a character vector of customer labels.

"adjust" is a non-negative number, an adjustment parameter for scaling the service times.

The package uses the "ggplot2" package (Wickham H., 2009) for the graphics. To plot the results, use the command "plot":

```
plot(x, which = c(2:6), annotated = TRUE,...)
```

Where:

"x" is an object of class queue_list

"which" is a numeric vector of integers from 1 to 7 which represents which plots are to be created.

"annotated" is a logical variable, if its value is TRUE, annotations will be added to the plot.

Example 8.7: An analysis is being made of a queuing system with several servers. Start with an arrival rate (λ) of 20 units an hour and c= 3 servers, each with a service rate (μ) equal to 8 units an hour. Performance measurements are required.

```
>library(queuecomputer)
>library(ggplot2)
>set.seed(123)
>n<-4000
>arrivals <- cumsum(rexp(n,20)) #cumsum() es un commando que
forma un vector con los valores acumulados del vector creado
>service <- rexp(n,8)
>queue_obj <- queue_step(arrivals,service= service, servers =
3)
```

```
>summary(queue_obj)
Total customers:
 4000
Missed customers:
 0
Mean waiting time:
 0.126
Mean response time:
 0.251
Utilization factor:
 0.833
Mean queue length:
 2.51
Mean number of customers in system:
 5
```

If the number of customers is changed to n = 10,000; you get:

```
> set.seed(123)
> n<-10000
> arrivals<-cumsum(rexp(n,20))
> service<-rexp(n,8)
> queue_obj <- queue_step(arrivals,service= service, servers =
3)
> summary(queue_obj)
Total customers:
 10000
Missed customers:
 0
Mean waiting time:
 0.156
Mean response time:
 0.282
Utilization factor:
 0.832
Mean queue length:
 3.11
Mean number of customers in system:
 5.61
Plotting the result:
> Plot(queue_obj,which=3)
```

Figure 7. Density for system time and waiting time, from running a simulation with n=10,000 customers. Using R package "queuecomputer"

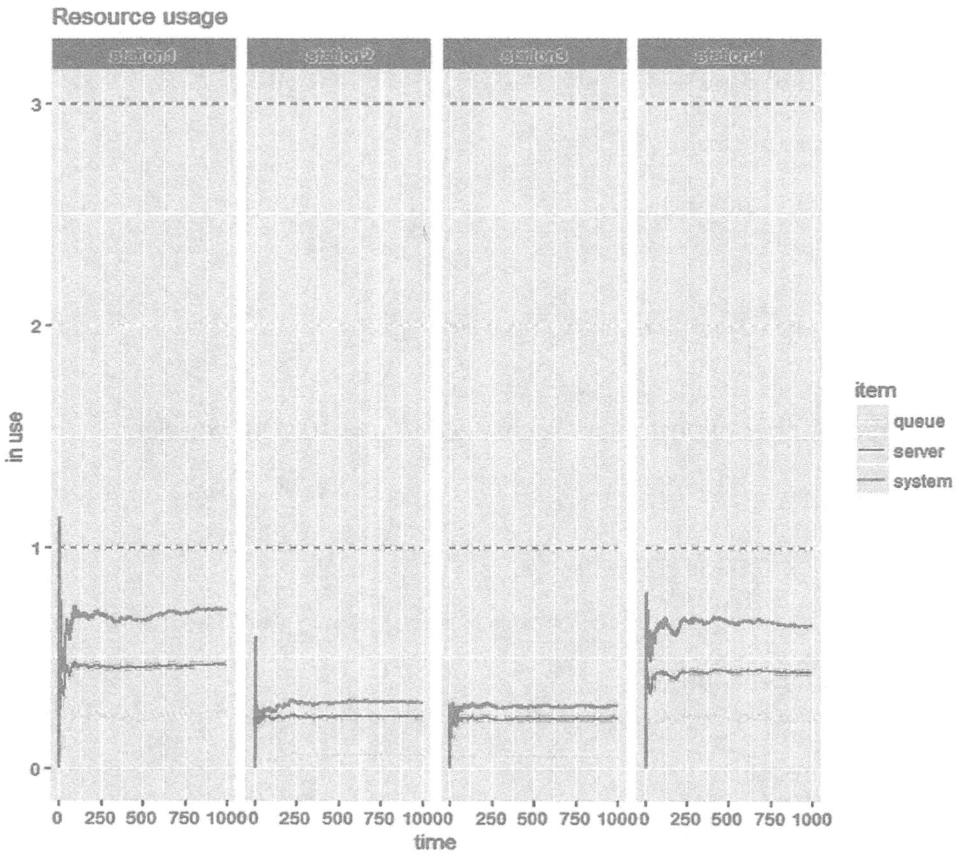

lag_step Function

The lag_step command adds a lag to a vector of arrival times. The lag_step command is equivalent to queue_step with a large number of queues, but is faster to calculate. The command format is:

```
lag_step(arrivals, service)
```

Where:

"arrivals" can be either a numeric vector or an object of class queue_list. It represents the arrival times.

"service" is a vector of service times with the same ordering as "arrivals".

In the end you get a vector of response times for inputs of arrival time and service times.

wait_step Function

The wait_step function calculates the maximum time for each row from two vectors of arrival times The command format is:

```
wait_step(arrivals, service)
```

Where:

"arrivals" can be either a numeric vector or an object of class queue_list. It represents the arrival times.

"service" is a vector of times that represents the arrival times of the second type of customers. The ordering of this vector corresponds to the same ordering as the arrivals vector.

Example 8.8: A Network - To show how the commands lag_step and wait_step can be applied, here is a variant of the example given in the help for the "queuecomputer" package for networks. You have the situation of passengers disembarking from a plane, then going through customs and finally going to pick up their baggage that has a different arrival time.

```
> library(queuecomputer)
> library(dplyr)
> set.seed(123)
> n<-160
```

Passenger arrival times (arrivals), the time involved in transferring from the plane to the arrivals hall (service_1), the time spent queueing for customs (service_q) and luggage arrival times (arrivals_b) are all defined.

```
> arrivals <- cumsum(rexp(n))
> service_1 <- rexp(n, 0.8)
> service_q <- rexp(n, 0.5)
> arrivals_b <- cumsum(rexp(n, 0.8))
```

The transfer departure times (departures_1), the times of departure from the arrivals procedure when there are two servers (departures_2) as well as the times it takes passengers to find their baggage (departures_3) are all calculated.

```
> departures_1 <- lag_step(arrivals, service_1)
> departures_2 <- queue(departures_1,service=service_q,serve
rs=2)
> departures_3 <- wait_step(departures_2, arrivals_b)
```

The results can be observed as:

```
summary(departures_3)
 Min. 1st Qu. Median Mean 3rd Qu. Max.
 4.609 51.290 97.820 97.460 142.200 193.700
> summary(arrivals)
 Min. 1st Qu. Median Mean 3rd Qu. Max.
 0.8435 38.8000 81.6600 81.3000 118.2000 157.3000
```

Plotting:

Figure 8. Plotting of the times for the 160 passengers. The arrival times marked with dots. The departure times with luggage form the top line. The departure times with arrival procedure form the middle line.

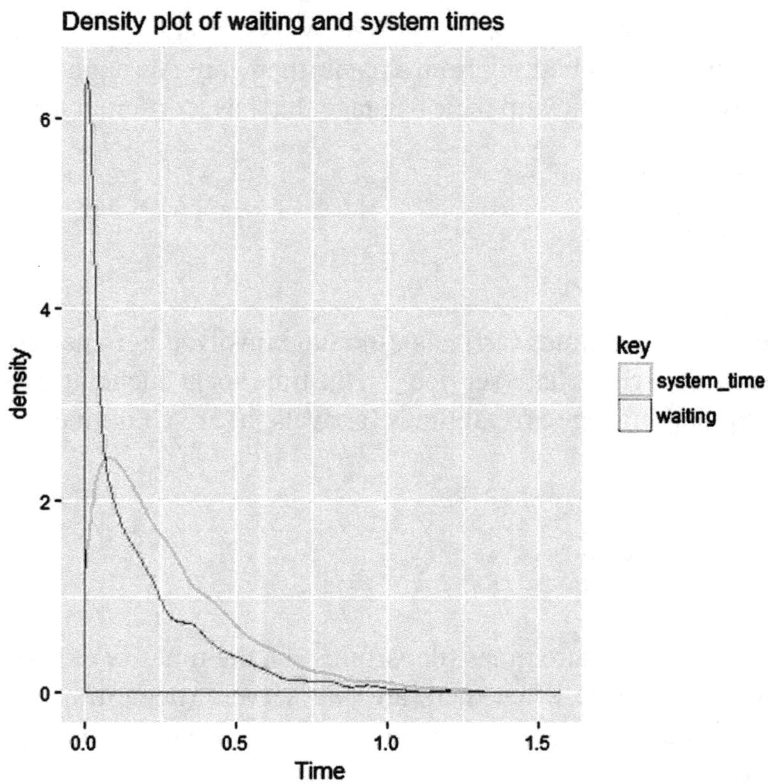

```
> plot(arrivals, main="Arrivals and departures times")
> lines(departures_3)
> lines(departures_2)
```

There are a variety of programs for the calculation and simulation of queueing systems. There is literature that has periodically presented lists of commercial software and a general list of free simulation software can also be found in the network.

Some considerations based on the analysis of the results of simulations on measurements are given

Descriptions and examples are presented of free software included as packages in the R Programming language, for the calculation of queueing systems. These programs have more extensive documentation and examples on the communication and documentation sites. We invite any reader who is interested to follow them.

REFERENCES

Bard, Y. (1979). Some extensions to multiclass queueing network analysis. *Proceedings of 4th International Symposium on Modelling and Performance Evaluation of Computer Systems*, 51–62.

Canadilla, P. (2016). *_queueing: Analysis of Queueing Networks and Models_*. R package version 0.2.8. Retrieved from https://CRAN.R-project.org/package=queueing

Chambers, J. M. (1998). *Programming with Data*. New York: Springer. doi:10.1007/978-1-4684-6306-4

Dagkakis, G., & Heavey, C. (2016). A review of open source discrete event simulation software for operations research. *Journal of Simulation*, *10*(3), 193–206. doi:10.1057/jos.2015.9

Davison, A. C., & Hinkley, D. V. (1997). *Bootstrap methods and their application* (Vol. 1). Cambridge University Press. doi:10.1017/CBO9780511802843

Dias, L. M., Vieira, A. A., Pereira, G. A., & Oliveira, J. A. (2016, December). Discrete simulation software ranking—A top list of the worldwide most popular and used tools. In *Winter Simulation Conference (WSC)* (pp. 1060-1071). IEEE. doi:10.1109/WSC.2016.7822165

Ebert, A. (2017). *queuecomputer: Computationally Efficient Queue Simulation*. R package version 0.8.1. Retrieved from https://CRAN.R-project.org/package=queuecomputer

Ebert, A., Wu, P., Mengersen, K., & Ruggeri, F. (2017). *Computationally Efficient Simulation of Queues: The R Package queuecomputer*. arXiv preprint arXiv:1703.02151

Faraway, J., Marsaglia, G. Marsaglia, J., & Baddeley, A. (2017). *goftest: Classical Goodness-of-Fit Tests for Univariate Distributions*. R package version 1.1-1. Retrieved from https://CRAN.R-project.org/package=goftest

Hlynka, M. (2014, June 10). *List of Queueing Theory Software*. Retrieved January 01, 2017, from http://www2.uwindsor.ca/~hlynka/qsoft.html

Law, A. M. (2015, December). Statistical analysis of simulation output data: the practical state of the art. In *Proceedings of the 2015 Winter Simulation Conference* (pp. 1810-1819). IEEE Press. doi:10.1109/WSC.2015.7408297

Mourtzis, D., Doukas, M., & Bernidaki, D. (2014). Simulation in manufacturing: Review and challenges. *Procedia CIRP*, *25*, 213–229. doi:10.1016/j.procir.2014.10.032

R Core Team. (2017). *R: A language and environment for statistical computing*. R Foundation for Statistical Computing. Retrieved from https://www.R-project.org/

Reiser, M., & Lavenberg, S. S. (1980). Mean-value analysis of closed multichain queueing networks. *Journal of the Association for Computing Machinery*, *27*(2), 313–322. doi:10.1145/322186.322195

Schweitzer, P. (1979). Approximate analysis of multiclass closed network of queues. *Proceedings of International Conference on Stochastic Control and Optimization*, 25–29.

Swain, J. J. (2015, October). *Simulation Software Survey*. Retrieved January 01, 2017, from OR/MS Today: https://www.informs.org/ORMS-Today/Public-Articles/October-Volume-42-Number-5/Simulation-Software-Survey

Swain, J. J. (2013). Discrete Event Simulation Software Tools: A better reality. *OR/MS Today*, *40*(5), 48-59.

Thiede, S., Seow, Y., Andersson, J., & Johansson, B. (2013). Environmental aspects in manufacturing system modelling and simulation—State of the art and research perspectives. *CIRP Journal of Manufacturing Science and Technology*, *6*(1), 78–87. doi:10.1016/j.cirpj.2012.10.004

Ucar, I., & Smeets, B. (2017). *simmer.plot: Plotting Methods for 'simmer'*. R package version 0.1.9. Retrieved from https://CRAN.R-project.org/package=simmer.plot

Ucar, I., & Smeets, B. (2017). *simmer: Discrete-Event Simulation for R.* R package version 3.6.2. Retrieved from https://CRAN.R-project.org/package=simmer

Ucar, I., Smeets, B., & Azcorra, A. (2017). *simmer: Discrete-Event Simulation for R.* arXiv preprint arXiv:1705.09746

Venables, W. N., & Ripley, B. D. (2013). *Modern applied statistics with S-PLUS*. Springer Science & Business Media.

Wand, M. P., & Jones, M. C. (1994). *Kernel smoothing*. CRC Press.

Wickham H. (2009). *ggplot2: Elegant Graphics for Data Analysis*. Springer-Verlag.

Glossary

Allen-Cunneen Approximation: Let us assume that demand follows a general probability distribution (*G*) with mean t_a and standard deviation σ_a and that the service discipline is First In, First Out and independent of the service time; let us also assume that for the service time, the only specification is for it to follow a general probability distribution (*G*) with mean t_S and standard deviation σ_S and for the station to have several parallel servers (c), Allen-Cuneen approximation for the time in queue in G/G/c systems is:

$$CT_q \approx \left(\frac{C_a^2 + C_S^2}{2}\right)\left(\frac{\rho^{\sqrt{2(c+1)}-1}}{(1-\rho)c}\right)t_S$$

where $C_S^2 = \frac{\sigma_S^2}{(t_S)^2}$; $C_a^2 = \frac{\sigma_a^2}{(t_a)^2}$

Anderson-Darling Test: The Anderson-Darling test is a nonparametric test that performs a statistical hypothesis test with the null hypothesis Ho: the data of a sample follow a distribution. The test is applied to univariate continuous random variables and variants of the test have been developed for different distributions. See *Goodness of fit tests.*

Arrival: The entry of a customer is called an arrival. The customers may arrive to request a service either randomly or at well-defined intervals of time (typical of an automated processes). customers arrive at the system with a certain periodicity (t_a). In queues, the number of customers that arrive per time period is represented by the Greek letter λ and $\lambda = 1/ t_a$.

Birth and Death Processes: A birth – death process is a class of Markov chains where time happens continuously. In these types of processes there is a set of states {0, 1, 2, …} that is called "population" and the transitions between states indicate increments or decrements in the number of individuals in the "population".

Bottleneck: In general, there is no formal definition of a bottleneck. The convention is that it is a station that limits or restricts the flow in a system.

Bottleneck, Economic: This is the station for which a change in the rate of service will generate the largest reduction in the cash flow cost; an analysis of the marginal cost enables us to find the station we should pay attention to from amongst a set of stations that form the network.

Burke's Theorem: Burke's Theorem says that, for a queueing system, with a single queue, without departures, for any number of servers and independent times between arrivals with exponential distribution, the time intervals between successive outputs are independently and exponentially distributed with a parameter equal to the parameter for the exponential distribution of times between arrivals.

Central Limit Theorem: The Central Limit Theorem states that if S_n is the sum of n independent random variables, then the distribution of S_n is an approximation of the normal distribution given by the density function:

$$f\left(x\right)=\frac{1}{\sqrt{2\pi\sigma}}e^{-\left(x-\mu\right)^{2}/\left(2\sigma^{2}\right)}$$

where *f(x)* is a normal probability density function.

Chi-Square Goodness-of-Fit Test: The chi-square goodness-of-fit test is a test for assessing the goodness of fit in the event that the data appear in the form of class marks and their frequencies, such as when there is a histogram

Coefficient of Variation: The coefficient of variation is the division by the square root of the variance, called the standard deviation, and the expected value of the random variable is equally defined as a discrete or a continuous variable: $C\left(X\right)=\dfrac{\sqrt{V\left[X\right]}}{E\left[X\right]}$. It is often more practical to use the coefficient of variation squared, which in this case is: $C^2\left[X\right]=\dfrac{V\left[X\right]}{E^2\left[X\right]}$.

Congestion: This is the measurement that gives the proportion of capacity being used for a particular demand for service. This is also known as traffic intensity and is represented by the Greek letter ρ (rho). The congestion value must be less than 1 for the system to be considered stable and if not, that is a sign that the system does not have enough capacity to meet the demand for service and consequently the line or queue will grow unceasingly.

Cost (Operating Cost of a Queueing Network): The operating cost of a network has the following general expression: Cost=E[ServiceCost]+ E[WaitingCost]. The service cost for a system with *n* stations is:

$$\mathrm{E}\left[\mathrm{ServiceCost}\right]=\sum_{i=1}^{n}C_{Service,i}\times c\times\mu_i$$

Assuming that the waiting cost is proportional to the waiting time:

$$\mathrm{E}\left[\mathrm{WaitingCost}\right]=\sum_{i=1}^{n}C_{WIPt,i}\times WIP_{S,i}\,.$$

Customer: This is the entity that asks for a service and should be served. The customers are sometimes not obvious or evident at first sight, for example, a passenger waiting at a bus stop or terminal.

Cycle Time in the Queue: Is the time that the customer is in a queue waiting for service.

Cycle Time in the System: Is the sum of the cycle time in the queue and the expected value of the service time. The cycle time in the system is the time a customer is there from the moment it enters, lines up and goes through the service period until it leaves the system.

Erlang Distribution: If we add "k" exponential random variables with the same parameter λ, the resulting sum will be a random variable with a distribution called Erlang distribution with parameters λ and k, the density function is:

$$f(t) = \lambda \frac{(\lambda t)^{k-1}}{(k-1)!} e^{-\lambda t} \text{ for } t > 0$$

The cumulative probability distribution function is:

$$F(t) = 1 - \sum_{j=0}^{k-1} \frac{(\lambda t)^{j}}{j!} e^{-\lambda t} \text{ for } t \geq 0$$

The expected value, variance and coefficient of variation are respectively E[T]= k/λ; V[T] = k/λ^2; C[T]= $1/\sqrt{k}$.

Expected Value: Expected value gives the value where the "center of gravity" of the probability distribution is. This is the point of equilibrium for distribution and is expressed as

$$E[X] = \int_{-\infty}^{\infty} x f(x)\, dx \text{ X continuous}$$

$$E[N] = \sum_{k=-\infty}^{\infty} k f(k) \text{N discrete}$$

Exponential Distribution: A non-negative variable T follows an exponential distribution with parameter λ, if its density function is: $f_T(t) = \lambda e^{-\lambda t}, t \geq 0$. With Cumulative distribution function: $F_T(t) = 1 - e^{-\lambda t}, t \geq 0$. The expected value, variance and coefficient of variation are respectively E[T]=1/λ; V[T]= 1/λ^2; C[T]=1. For the analysis of stochastic processes in general and queues in particular, exponential distribution has a very important property, which is the absence of memory.

Failures on Cycle Time: In the examples given so far, the equipment or stations do not break down and are always available, this is an ideal situation because equipment fails in real life, causing interruptions of production and consequently lowering the equipment's availability. The availability is a dimensionless number $a \in (0,1)$ is calculated using the following expression:

$$a = \frac{MTTF}{MTTF + MTTR}$$

where *MTTR* is the average repair time, *MTTF* is the average operating time until the catastrophic failure occurs. The availability of the equipment can be used to calculate the effective time of the station and the variance in the effective time:

$$t_e = \frac{t_S}{a}$$

$$\sigma_e^2 = \left(\frac{\sigma_S^2}{a}\right) + \frac{\left(MTTR^2 + \sigma_{MTTR}^2\right)\left(1-a\right)}{aMTTR} t_S$$

where σ_e^2 is the effective variance; t_e is the effective service time; t_S the service time; σ_S^2 the variance in the service time and σ_{MTTR}^2 the variance in the average repair time.

Goodness of Fit Tests: Goodness of Fit Tests test the hypothesis: these data follow the given probability distribution. Let us assume that we have a sample of *n* observed random data for a process or population: $x_1, x_2, \ldots, x_n$ and we want to prove or find out what probability distribution model is the most suitable for describing them.

Jackson's Theorem: Let us assume that the service time and the time between consecutive customers follow an exponential distribution. Let k_i be the number of customers in station *i* and the vector that defines the state of the system in a given instant be $(k_1, k_2, \ldots, k_n)$. The probability P_k^i of finding *k* customers in station *i* is obtained by the following expression:

$$P_k^i = \begin{cases} P_0^i \dfrac{\left(\frac{\gamma_i}{\mu_i}\right)^k}{k!} & k = 0,1,\ldots,m_n \\ P_0^i \dfrac{\left(\frac{\gamma_i}{\mu_i}\right)^k}{m!}\left(m\right)^{k-m_n} & k = m_n, m_{n+1},\ldots \end{cases}$$

Based on above expression, the probability distribution of the states of the system described by the above expression has the following form:

$$P_k^i = \left(P_{k_1}^1\right)\left(P_{k_2}^2\right)\left(P_{k_3}^3\right)\ldots\left(P_{k_i}^i\right)$$

The model establishes that if the system is in a stationary state, the stations that form the network will behave like systems that are independent of each other and the system of equations for the flow of customers in the network (also known as traffic equations)

$$\boldsymbol{\lambda} = \left(\boldsymbol{I} - \boldsymbol{P}\right)^{-1}\gamma$$

satisfies the expression

$$P_k^i = \left(P_{k_1}^1\right)\left(P_{k_2}^2\right)\left(P_{k_3}^3\right)\ldots\left(P_{k_i}^i\right),$$

where $\boldsymbol{\gamma}$ is the external flow vector and a known value; $\boldsymbol{\lambda}$ is the flow vector for inputs to the stations that make up the network that has to be calculated; and $\mathbf{P}$ is the probability matrix or process path matrix and is known. Jackson's theorem establishes that for networks with M/M/1 or M/M/c stations, the entry of customers into each station follows a Poisson distribution and is not invalidated even in the case of a reflow or return to a station.

Kendall's Notation: In 1953, David George Kendall designed a notation that allows us to classify a queue according to the characteristics of the time that elapses between consecutive customers arriving in the system, the characteristics of the station's service time and the number of servers. This notation has been amplified to include the service discipline and the number of spaces available for the customers in the system:

```
Arrives/Service/Number of servers/Discipline/Capacity
```

The first symbol corresponds to the probability distribution for the time that elapses between two consecutive customers. The second symbol is the probability distribution for the service time. The third symbol is an integer number for identifying the number of servers. The fourth symbol is for the rule or discipline for customer service. And finally, the last symbol is used to define the number of places available for customers including the ones being served.

Kingman's Approximation of Cycle Time: Let us assume that demand follows a general probability distribution (*G*) with mean t_a and standard deviation σ_a and that the service discipline is First In, First Out and independent of the service time; let us also assume that for the service time, the only specification is for it to follow a general probability distribution (*G*) with mean t_S and standard deviation σ_S and for the station to have a single server John Frank Charles Kingman developed the following formula to approximate the time in queue of a G/G/1 system where $\rho = t_S / t_a$ is very close to 1:

$$CT_q \approx \left(\frac{C_a^2 + C_S^2}{2}\right)\left(\frac{\rho}{1-\rho}\right)t_S$$

where: $C_S^2 = \dfrac{\sigma_S^2}{\left(t_S\right)^2}$; $C_a^2 = \dfrac{\sigma_a^2}{\left(t_a\right)^2}$.

Kolmogorov-Smirnov Test: The Kolmogorov-Smirnov test is a non-parametric test that lets us compare two cumulative probability distributions for continuous variables. A statistic called D is calculated that estimates the longer vertical distance between the two cumulative distribution functions and if the statistic exceeds a critical value obtained for the distribution of distances, the null hypothesis that both distribution functions are equal is rejected. See *Goodness of fit tests.*

Lack of Memory: It turns out that the conditional probability of an event of interest being observed in *s* instants of time, knowing that *t* has already past, is independent of the time *t* that has past. This is posed as:

$$P\left(T > t + s \middle| T > t\right) = P(T > s).$$

Geometric and Exponential probability distributions have this property.

Little's Law: This is an equation that was developed and tested by D.C. Little in 1961 and links the amount of work in process to demand and cycle time. The equation is as follows: TC= WIP/ λ. Equation indicates that the average cycle time, TC, is obtained by dividing the amount of average work in process, WIP, by the average demand λ (customers arriving per unit of time).

M/M/1 System: The interarrival time and service time follow an exponential distribution, then this is a particular case of Markov chains. the customers arrive at a single-server queue with time intervals of $1/\lambda$ units of time per customer and the server provides the service at a speed of μ customers per unit of time.

Congestion: $\rho = \lambda / \mu$;

Work in system: $WIP_S = \dfrac{\rho}{1-\rho}$;

Work in queue: $WIP_q = \dfrac{\rho^2}{1-\rho}$

Time in system: $CT_S = \dfrac{1}{\mu - \lambda}$;

Time in queue: $CT_q = \dfrac{\rho}{\mu - \lambda}$

See: Single-Stage, Single Server Queueing Systems, Kendall's Notation.

M/M/1/K System: The interarrival time and service time follow an exponential distribution, then this is a particular case of Markov chains. the customers arrive at a single-server queue with time intervals of $1/\lambda$ units of time per customer and the server provides the service at a speed of μ customers per unit of time. the space for the customers is now limited to a certain number of spaces K (1 receiving service and the rest in a queue). The probability of all the spaces in the system being occupied $n = K$ while $\rho < 1$ is:

$$p_K = \frac{1-\rho}{1-\rho^{K+1}}\rho^K$$

In the fraction of time that the system permits entry $1 - p_K$, those customers for which there is space in the system are served, then the expression for the customers that were served is: $\lambda_{out} = \lambda_{in}\left(1 - p_K\right)$

The amount of work in system is:

$$WIP_S = \frac{\rho\left[1-\left(K+1\right)\rho^K + K\rho^{K+1}\right]}{\left(1-\rho\right)\left(1-\rho^{K+1}\right)} \qquad \rho \neq 1$$

For Work in queue:

$$WIP_q = WIP_S - \lambda_{out}/\mu \quad \rho \neq 1$$

The time a customer stays in the system only applies to customers that have been served and is obtained by using Little's Law:

$$CT_S = \frac{WIP_S}{\lambda_{out}} = \frac{WIP_S}{\lambda\left(1 - p_K\right)}$$

the time in queue:

$$CT_q = \frac{WIP_S}{\lambda\left(1 - p_K\right)} - \mathrm{E}\left[t_S\right].$$

M/M/c System: Is a queueing system with the time between consecutive customers is a random variable and follows an exponential distribution. The system now has c servers that operate independently, while the service time is also a random variable and follows an exponential distribution. Let us also assume that there is no difference between servers.

Congestion: ρ = $\lambda / c\,\mu$; the probability of the system being empty:

$$p_0 = \left(\frac{\left(\lambda/\mu\right)^c}{c!\left(1-p\right)} + \sum_{n=0}^{c-1} \frac{\left(\lambda/\mu\right)^n}{n!} \right)^{-1};$$

Work in queue:

$$WIP_q = \frac{\left(\lambda/\mu\right)^{c+1}}{\left(c-1\right)!\left(c-\lambda/\mu\right)^2} p_0;$$

Work in system:

$$WIP_S = WIP_q + \frac{\lambda}{\mu};$$

The time in system is obtained by using Little's Law: $CT = WIP/\lambda$.

Markov Chain: Is a stochastic process where, when a system is found in a particular state, the probability of getting the next action or result is not affected by the events or the path followed in the instants before the current state. A Markov chain (or process) $X(t)$ is a stochastic process with the property that if an event X_t arises, then all other possible future events X_t for the following instants of time $s > t$ are not affected by any past actions or events $u < t$ that brought the system to state X_t.

Markov Matrix: Also called Transition Matrix (P). The matrix is a matrix square with a finite number of rows and columns equal to the number of states of the system. Each line has the following properties:

$p_{ij} > 0$ for $i, j = 0, 1, 2, \ldots$; $\sum_{i=0}^{\infty} p_{ij} = 1$, $i = 0, 1, 2, \ldots$

A Markov process is defined by its transition matrix *P* and the initial state X_0. Each entry p_{ij} of the matrix indicates the probability of going from one state to another.

Model: Models are representations of systems and describe the characteristics that are of interest for the decisions that the administrator has to make. The process of creating a model is known as *modeling*.

Model, Symbolic: A symbolic model shows the relationships between each component of a system through the use of symbols or operators. The best-known ones are mathematical models.

Modeling Stages: Mathematical models are built out of the need to represent a system; in the field of decision-making, mathematical models are used to speed up decision-making and reduce the uncertainty associated with the effect or result of taking an action. A series of stages or phases are named: Defining problem, Modeling the System, Collecting data, Validation, Obtain a solution, Analyzing results, and Implementing solution.

Network Layout, Queueing Systems: Sometimes customers or pieces that are in the process follow a specific route within the process, because there are several options for paths to be followed. When these paths are represented, the system looks like a network with the stations interconnected.

Networks, Queueing: For the analysis of queueing networks, let us assume that: 1) The system has *n* work stations. Each station has 1 or more servers and there is no difference between them; 2) Customers enter the system through any one of the stations; 3) The service time in each station is t_s; 4) Once the service is finished, the customer leaves station *i* and goes on to the next stage *j* with probability p_{ij}, it can even leave the system through any one of the stations if process is completed; and 5) There is no constraint on the number of customers that can stay in the queue. The input and output flow balance equations show that the number of customers that circulate within a network is found by using the following equation:

$$\lambda_i = \gamma_i + \sum_{j=}^{n} \lambda_j p_{ji}; i = 1,2,\ldots,n$$

The above Equation establishes that the total input flow to a station i is the sum of the external flow plus proportion p_{ij} of the flow that leaves station j heading for station i. The solution of this system of linear equations is obtained using any known method for this case. For example, the above system has the following representation in matrix form:

$$\boldsymbol{\lambda} = \gamma + \boldsymbol{\lambda P}$$

where $\boldsymbol{\gamma}$ is the external flow vector and a known value; $\boldsymbol{\lambda}$ is the flow vector for inputs to the stations that make up the network that has to be calculated; and **P** is the probability matrix or process path matrix and is known. The system has the following solution:

$$\boldsymbol{\lambda} = \left(\boldsymbol{I} - \boldsymbol{P}\right)^{-1} \gamma$$

where **I** is an identity matrix of the same size as **P.**

Performance Measures: Queueing systems are analyzed by means of equations that make it possible to have a set of performance measures that can give us an idea about how they operate. The basic performance measures for queues are the cycle time, work in process as well as the congestion, and a set of equations have been developed that give information about the relationship between the elements of a queueing system and the impact they have on the time a customer stays in the system or on the number of pieces waiting for their turn to be processed.

Poisson Distribution: If N is defined as a random variable that counts the number of events of interest in a time interval. We get a density function, for a value "n" that tends to grow a lot and an average constant value λ:

$$f\left(N = k\right) = \lim_{n\to\infty}\left[\frac{n!}{k!\left(n-k\right)!}\left(\frac{\lambda}{n}\right)^{k}\left(1-\frac{\lambda}{n}\right)^{n-k}\right] = \frac{\lambda^{k}e^{-\lambda}}{k!} \text{ for } k = 0,1,\cdots$$

where: The function between the square brackets is the binomial probability function. The number "e" is the so-called natural number. "e" is the basis for natural logarithms, approximately 2.718281828459.

The resulting probability density function is called Poisson distribution. The expected value, variance and coefficient of variation of the Poisson distribution are:

$$E[k] = \lambda,\ V[k] = \lambda,\ C[k] = \frac{1}{\sqrt{\lambda}}$$

Pollacheck-Kintchine Approximation for the Cycle Time: Let us assume that demand λ follows a Poisson distribution and that the service discipline is First In, First Out and independent of the service time; let us also assume that for the service time, the only specification is for it to follow a general probability distribution (G) with mean t_S and standard deviation σ_S and for the station to have a single server. Under Kendall's notation, these types of systems are M/G/1 and M/G/c. The time in queue is obtained using the following expression:

$$CT_q = \left(\frac{1 + C_S^2}{2}\right)\left(\frac{\lambda t_S}{1 - \lambda t_S}\right) t_S$$

where the parameter $C_S^2 = \frac{\sigma_S^2}{\left(t_S\right)^2}$ is the squared coefficient of variation. The equation is known as a Pollaczek-Kintchine Formula and, just like the Markov systems, the system has to be stable, in other words, $\lambda t_S = \rho \le 1$. The amount of work in system is:

$$WIP_S = \left(\frac{1 + C_S^2}{2}\right)\left(\frac{\rho^2}{1 - \rho}\right) + \rho .$$

Quantile: Assuming that we have a cumulative probability value, p = F(x), for a numeric value x, the value of x for the random variable that corresponds to that cumulative probability is called quantile (p). A quantile) or q(p), complies with:

$$p = F_X\left(quantile\left(p\right)\right) = \int_{-\infty}^{quantile(p)} f\left(x\right) dx \text{ X continuous}$$

If p=F(x), then q(p)= $F^{-1}(p)$= x.

Quantitative Analysis: Is defined as the scientific approach to problems or conflictive situations in a system, where the administrator has to make a decision.

Queue: This is made up by to be customers waiting to be served. The length of a queue makes a strong visual impact and people make a direct connection between its size and the quality of service, seeing it as the time it takes to be served.

Queueing System, Elements: A Queueing system, that is made up by a set of elements: customers, server, queue and demand.

Queueing Theory: The waiting phenomenon is studied by a specialist field called Queueing Theory, this field has developed a set of models that enable us to understand the causes of waiting, measure its impact in quantitative terms and analyze different alternatives or options speed up the flow of customers through a system. Waiting to be seen to is a phenomenon that is inherent to processes and basically its perception is based on time.

R: Is a language and environment for statistical computing and graphics. R provides a wide variety of statistical (linear and nonlinear modelling, classical statistical tests, time-series analysis, classification, clustering, …) and graphical techniques, and is extensible. R is available as Free Software under the terms of the Free Software Foundation's GNU General Public License in source code form. It compiles and runs on a wide variety of UNIX platforms and similar systems (including FreeBSD and Linux), Windows and MacOS.

Scilab: Is free and open source software for numerical computation providing a computing environment for engineering and scientific applications. Scilab is released as open source under the CeCILL license (GPL compatible), and is available for download free of charge. Scilab is available under GNU/Linux, Mac OS X and Windows XP/Vista/7/8.

Series, Queueing Systems: In this case the complete service requires several phases or stages. The route is the same for each customer and assumes that entry is exclusively through the first stage or station of the process.

Server: This is the person who serves the customer a coffee and charges for it; the machine that processes the piece; or the machine that collects parking fees. Queueing systems may have a single server or several parallel servers. The server requires a certain time to execute a task (the service or manufacturing process) and this usually varies a little from customer and customer.

Service Capacity: The service capacity of a queueing system is the number of customers that can be served during a certain period of time (t_s). Generally, in queues, the average number of customers served per unit of time is used as a measure of the capacity of a server and represented by the Greek letter μ (mu). Service capacity is in inverse proportion to service time.

Service Discipline: This refers to the order or priority in which customers or users are served. The most common one is First In, First Out.

Single-Stage, Parallel Servers Queueing Systems: The customer enters the system where they wait to be served, then the service is completed in just one stage, however in this case there are a variety of servers serving several customers at the same time. Once the server has finished serving a customer, it is free to receive the next waiting customer.

Single-Stage, Single Server Queueing Systems: The customer enters the system where they wait to be served, then as soon as a server is free, the customer goes in be served. When the service finishes, the customer leaves the system and the next customer goes in.

Stochastic Process: A stochastic process is defined by: 1. The possible events of set of random variables *X*, which are known as a set of possible states or just as states, $X(t)$ with a parameter *t*, whose elements belong to a set *T*, 2. The elements of set *T*, 3. The relationship or connection that exists between every possible state. In applications of stochastic processes for modeling real systems it is very common for set *T* to refer to a time sequence and each element *t* of the set to represent (discrete or continuous) instants of time within the same sequence.

Systems: Systems consist of a variety of components or elements. Each element is connected to the others and together they form an entity that produces an answer. Systems, in decision-making, are those entities that are under the responsibility of an administrator who has to make a series of decisions that should generate the answer needed to achieve an objective.

Systems in Series With G/G/c Stations: In the case where the service time and time between arrivals follow general distributions, we need to resort to approximations in order to get the properties. The flow that leaves a station becomes the demand flow of the next station, so the output variability needs to be calculated as this will be the input variability in the next stage of the process. Let us assume that the service time follows a probability distribution with a mean of $1/\mu$ and a standard deviation of σ. Likewise let us assume that the time between arrivals follows a probability distribution with a mean of $1/\lambda$ and a standard deviation of σ. The cycle time and squared coefficient of variation of the output from station 1 are calculated using the expressions:

$$CT_{S,1} \approx \left(\frac{C_a^2 + C_{S,1}^2}{2}\right)\left(\frac{\rho_1^{\sqrt{2(c_1+1)}-1}}{(1-\rho_1)c_1}\right)t_{S,1} + t_{S,1}$$

$$C_{d,1}^2 \approx C_a^2 + 2\rho_1^2 C_{S,1}^2 - \rho_1\left(\frac{C_a^2 + C_{S,1}^2}{2}\right)\left(\frac{\rho_1^{\sqrt{2(c_1+1)}-1}}{c_1}\right)$$

In this case, C_a^2 refers to the squared coefficient of variation of the flow from outside that enters through station 1. The flow of customers that leaves station 1 goes to station 2. The time between customers will be $1/\lambda$ and the variability at output C_{d1} will now be the variability with which the customers enter station 2:

$$CT_{S,2} \approx \left(\frac{C_{d,1}^2 + C_{S,2}^2}{2}\right)\left(\frac{\rho_2^{\sqrt{2(c_2+1)}-1}}{(1-\rho_2)c_2}\right)t_{S,2} + t_{S,2}$$

$$C_{d,2}^2 \approx C_{d,1}^2 + 2\rho_2^2 C_{S,2}^2 - \rho_2\left(\frac{C_{d,1}^2 + C_{S,2}^2}{2}\right)\left(\frac{\rho_2^{\sqrt{2(c_2+1)}-1}}{c_2}\right)$$

In these expressions, the value of C_a^2 is swapped for $C_{d,1}$. Given that all the customers that enter station 1 continue their path towards the station 2, then $\rho_2 = \lambda t_{s2}$. After station 2 the squared coefficient of variation of the output from previous station is used

After station 2 the calculation process is repeated for each one of the n stations that make up the line and can be generalized in the following recursive expressions and:

$$CT_{S,i} \approx \left(\frac{C_{d,i-1}^2 + C_{S,i}^2}{2}\right)\left(\frac{\rho_i^{\sqrt{2(c_i+1)}-1}}{(1-\rho_i)c_i}\right)t_{S,i} + t_{S,i}$$

$$C_{d,i}^2 \approx C_{d,i-1}^2 + 2\rho_i^2 C_{S,i}^2 - \rho_i\left(\frac{C_{d,i-1}^2 + C_{S,i}^2}{2}\right)\left(\frac{\rho_i^{\sqrt{2(c_i+1)}-1}}{c_i}\right)$$

where $C^2_{d,i-1}$ is the squared coefficient of variation of the flow from outside that goes to station i. Once the properties of station 1 are obtained, the information of C^2_d is used to calculate the cycle time of the next station.

Uniform Distribution: A uniform distribution with a simple structure for continuous variables that is very useful for simulating random events. It is defined as the probability function that assigns uniform probability values to a continuous random variable X, within a [a, b] interval:

$$f(x) = \frac{1}{b-a} a \leq x \leq b$$

where: f(x) is the probability density function; "a" is the lower bound of the interval for X, while "b" is the upper bound.

The corresponding cumulative distribution function is:

$$F(x) = P(X \leq x) = \frac{x-a}{b-a} a \leq x \leq b$$

where: F(x) is the cumulative distribution function; “x” the numeric value for which the probability is calculated, “a” and “b” are respectively the upper and lower bounds for the variable.

The expected value, variance and coefficient of variation are respectively:

$$E[X] = \frac{a+b}{2}; V[X] = \frac{(b-a)^2}{12}; C[X] = \frac{b-a}{\sqrt{3}(a+b)}.$$

Variance: Variance is a measure of dispersion of the random variable distribution and is defined as:

$$V[X] = \int_{-\infty}^{\infty} (x - E[X])^2 f(x)dx \text{ X continuous}$$
$$V[N] = \sum_{k=-\infty}^{\infty} (k - E[N])^2 f(k) \text{N discrete}$$

Work in Process in the Queue: Refers to customers that are exclusively lined up waiting for service (WIP_q).

Work in Process in the System: (WIP_S) is the sum of the customers lined up and the ones being served.

Index

www.ingramcontent.com/pod-product-compliance
Lightning Source LLC
LaVergne TN
LVHW061543290926
849162LV00004B/887